Genetic Engineering 2

Avtar Handa
June 7, 1982

村上功
Isao Murakami
岡山県倉敷市
大島175-4

175-4 Ohjima
Kurashiki-City
Okayama-Pref.
Japan

1982 Aug. 4

Genetic Engineering 2

Edited by

Robert Williamson

Professor of Biochemistry,
St Mary's Hospital Medical School,
University of London

ACADEMIC PRESS · 1981

A Subsidiary of Harcourt Brace Jovanovich, Publishers

London · New York · Toronto · Sydney · San Francisco

ACADEMIC PRESS INC. (LONDON) LTD
24/28 Oval Road,
London NW1

United States Edition published by
ACADEMIC PRESS INC.
111 Fifth Avenue,
New York, New York 10003

British Library Cataloguing in Publication Data

Genetic engineering.
Vol. 2
1. Genetic engineering — Periodicals
I. Williamson, R
575.1′05 QH 442

ISBN 0–12–270302–2

LCCCN 80–41976

Printed in Great Britain at the Alden Press
Oxford London and Northampton

Contributors

J.D. Beggs *Cancer Research Campaign, Eukaryotic Molecular Genetics Research Group, Department of Biochemistry, Imperial College, London SW7 2AZ, UK*

H.H. Dahl *Laboratory of Gene Structure and Expression, National Institute for Medical Research, The Ridgeway, Mill Hill, London NW7 1AA, UK*

R.A. Flavell *Laboratory of Gene Structure and Expression, National Institute for Medical Research, The Ridgeway, Mill Hill, London NW7 1AA, UK*

F.G. Grosveld *Laboratory of Gene Structure and Expression, National Institute for Medical Research, The Ridgeway, Mill Hill, London NW7 1AA, UK*

A.J. Jeffreys *Genetics Department, University of Leicester, Leicester LE1 7RH, UK*

A.D.B. Malcolm *Biochemistry Department, St Mary's Hospital Medical School, University of London, London W2, UK*

Preface

In the preface to Volume 1 of this series, I commented on the revolutionary nature of genetic engineering, and the fact that it permits qualitatively different kinds of scientific results to be obtained. The article in this volume on gene evolution illustrates this: Alec Jeffreys describes the study of divergence of coding and non-coding DNA sequences for related species, and shows how fossil evidence, protein sequence data and the study of genes using recombinants all tie together to give new evolutionary concepts. Dick Flavell and his colleagues outline the way in which genomic libraries can be made and analysed, a companion article for that from Jeff Williams in Volume 1, which may prove particularly interesting to those searching for particular human genes implicated in hereditary disease. Alan Malcolm outlines various considerations for the user of restriction enzymes: as many molecular biologists have no enzymological training, and regard enzymes (correctly) primarily as commercial products of remarkably high purity and reliability that are bought from suppliers, I think it particularly apt that such a clear summary should be available pointing out that there are basic enzymological properties of restriction endonucleases whose study leads to better experimental results. Finally, Jean Beggs gives a brief outline of new vectors available which allow cloning in yeast, a host which may have great commercial usefulness, as well as representing a stepping stone between prokaryotes and higher plants and animals.

It has again been a pleasure to edit this volume: the contributions will, in my view, be of interest both to the dedicated molecular biologist and the good student (of any age) hoping to learn more about the field. My colleagues at St Mary's, and at Academic Press, have been unfailingly helpful. However, the greatest pleasure has been the number of unsolicited positive comments from colleagues complimenting Volume 1. Whether this series meets a need can only be assessed in terms of its usefulness to those working in, and studying, recombinant DNA technology.

London, 28 April 1981 *Bob Williamson*

Contents

The use of genomic libraries for the isolation and study of eukaryotic genes

H.H. Dahl, R.A. Flavell and F.G. Grosveld

The use of restriction enzymes in genetic engineering

A.D.B. Malcolm

Gene cloning in yeast

J.D. Beggs

Recent studies of gene evolution using recombinant DNA

A. J. JEFFREYS

Genetics Department, University of Leicester, Leicester, UK

I Introduction

By comparing gene sequences, we can learn something of the mechanisms by which DNA evolves. This review will survey some recent advances in our understanding of the evolution of animal genes, which have resulted from the application of recombinant DNA methods to the analysis of gene structure and function.

A Setting the scene: the molecular evolution of proteins

The foundations of the study of molecular evolution were established by comparing amino acid sequences of related proteins (Dayhoff, 1972). Homologous protein sequences in different species were found to differ in a phylogenetically consistent fashion: the more closely related the species, the more similar the amino acid sequences. By comparing sequences, detailed molecular phylogenies were derived that in general reflected the evolutionary relationship between species as deduced from taxonomic and palaeontological studies. Most significantly, individual proteins were found to change in sequence during evolution at a constant rate, irrespective of the particular lineage being studied (see Wilson *et al.*, 1977). This evolutionary molecular clock runs at very different rates for different proteins, with a 400-fold difference between the highly conserved histone H4 and the rapidly evolving fibrinopeptide B.

Amino acid substitutions do not occur at random positions within polypeptides; for example, a number of residues in haemoglobin, particularly those associated with the haem binding site, are important for function and tend to remain invariant during evolution. Similarly, the difference in evolutionary clock rates between histones and fibrinopeptides suggests that there are relatively very few histone residues which can be replaced without affecting polypeptide function.

Amino acid sequencing has also revealed the existence of families of related proteins within a single species. For example, there are clear homologies between the α-, β-, γ-, δ- and ε-globin polypeptides of man, sufficient to establish that the corresponding globin genes must have arisen by some process of sequential gene duplication from an ancestral globin gene. By using clock rates for globin sequence divergence, it is possible to estimate the times at which these gene duplications occurred. For example, the first duplication giving rise to the α- and β-globin genes probably occurred some 500 million years ago, consistent with the existence of α- and β-globin in all higher vertebrates but not in primitive chordates such as the lamprey (Dayhoff, 1972).

These phylogenetic analyses have been complemented by studies of genetic variation of polypeptide sequences within a species (see Harris and Hopkinson, 1972; Harris, 1980). Both polymorphic and rare variants exist at many loci, and can be regarded as a major reservoir of genetic variability, which, through natural selection or genetic drift, can lead to the fixation of new primary polypeptide sequences in a population. Much controversy exists over the mechanisms responsible for maintaining this variability. The selectionist school argues that most or all polymorphisms are maintained by selection through, for example, heterozygous advantage, whereas neutralists argue that most variants are selectively neutral and attain polymorphic frequencies or become fixed by a process of random genetic drift. Similarly, neutralists would maintain that the clock-like evolution of polypeptides is the consequence of a constant rate of appearance and fixation, by drift, of selectively neutral protein variants. The selectionist argument is that amino acid substitutions are adaptive and that the clock-like mode of evolution reflects a species' response to a constantly changing environment (see Wilson *et al.*, 1977; Gale, 1980; Harris, 1980).

B Why study molecular evolution at the DNA level?

Little of the DNA in higher eukaryotes is used for coding proteins; the function of the remaining DNA, including intervening sequences, gene spacers and satellite DNAs is enigmatic (see Orgel and Crick, 1980). Possible genetic functions for these non-coding elements and the evolution of these functions can only be studied by analysing the molecular evolution of DNA. In particular, it is possible that major changes in morphology during evolution are the result of alterations not of proteins, but of extragenic regulatory elements (King and Wilson, 1975). Clearly, the analysis of protein evolution tells only part of the story, and must be extended by studies on DNA.

C Early studies on the evolution of DNA

Before the advent of recombinant DNA technology, evolutionary studies were confined to the entire genome or to readily isolated DNA subfractions (see Lewin, 1974). The first comparisons of genome size (C value) revealed that even closely related species can have widely differing C values; the mechanisms and evolutionary significance of these rapid genome expansions and contractions are still unknown. Repetitive and unique DNA sequences from different species were compared in more detail by DNA annealing. As with proteins, DNA sequence divergence could be used to construct species phylogenies, although there is some doubt whether nuclear DNA sequences evolve in a clock-like fashion (Kohne, 1970; Bonner *et al.*, 1980). The next major advance in evolutionary studies came with detailed analyses of repetitive DNA sequences, including apparently non-coding DNA sequences (such as satellite DNAs) and coding DNA (such as histone genes and ribosomal DNA) (see Hood *et al.*, 1975; Tartof, 1975; Kedes, 1979; Long and Dawid, 1980). Many of these sequences are arranged in tandem repetitive families which have probably arisen, in part at least, by tandem gene duplication. Expansion and contraction of the size of these families by unequal crossing over between homologous repeats was postulated as a mechanism for maintaining sequence homogeneity within a family. Other mechanisms, such as transposition or chromosomal translocation, were invoked to account for the dispersal of some repetitive sequence families (Tartof, 1975).

D The new studies: evolution of single copy genes

Within the last few years recombinant DNA technology has allowed us to detect and clone single gene sequences from the DNA of higher organisms, and has opened the way to a detailed analysis of the evolution of single genes and their associated DNA sequences. Although the field is still in its infancy, major advances have already been made in analysing the organization and evolution of gene clusters and in using phylogenetic comparisons to determine rates and modes of DNA sequence divergence within and outside genes. The discovery of intervening sequences within genes has led to some fascinating speculation on the evolution of gene structure in eukaryotes. Non-functional pseudogenes have been discovered, and apparently represent the evolutionary relics of once active genes. Parallel or "concerted" evolution of gene pairs has been shown to occur as a result of DNA sequence exchange between duplicated loci. No doubt many other evolutionary surprises are awaiting us as the analysis of single copy genes continues.

E Summary

Comparisons of genes isolated by recombinant DNA techniques are opening up a major new field of research in molecular evolution. Although these studies are an extension of previous phylogenetic analyses of protein sequences, total nuclear DNA and repetitive DNA sequences, they are showing much more precisely how genes have evolved and how new genetic functions are generated in evolution. This review will be concerned primarily with recent studies on the evolution of single copy animal genes and gene clusters, made possible by recombinant DNA methods. Earlier studies on the molecular evolution of proteins and repetitive DNA sequences have been fully reviewed elsewhere (Hood *et al.*, 1975; Tartof, 1975; Wilson *et al.*, 1977; Kedes, 1979; Long and Dawid, 1980).

II Experimental and theoretical approaches

A Detection and isolation of related genes

Most single copy genes examined to date have been detected using a cloned complementary DNA (cDNA) made from the required messenger RNA (see J. G. Williams, this series, Vol. 1). Labelled, cloned cDNA can be used to detect the corresponding homologous gene in restriction endonuclease digests of total genomic DNA, by agarose gel electrophoresis of DNA fragments followed by Southern blotting and filter hybridization (Southern, 1975, 1980). Similarly, cloned cDNA can be used to screen a recombinant λ bacteriophage-genomic DNA library, and to identify a recombinant plaque containing the corresponding gene (see R. A. Flavell, this volume).

Cloned cDNAs or genes can also be used to detect related gene sequences both in Southern blot hybridizations and in recombinant phage library screens. Detection will depend on the stringency of hybridization and the degree of DNA sequence divergence between the probe and the gene. For example, human β-globin cDNA detects the closely related β- and δ-globin genes at high stringencies of hybridization, plus the less closely related fetal globin genes ($^{G}\gamma$ and $^{A}\gamma$) at low stringencies (Flavell *et al.*, 1978). The maximum DNA sequence divergence between probe and gene compatible with detection of the gene in Southern blot analyses is about 25—30% over the region of greatest sequence homology. Somewhat greater levels of mismatch may be tolerated in plaque screenings of recombinant phage libraries, since the relative concentration of the relevant gene sequence in a plaque is greater than in genomic DNA on a Southern blot filter.

Cloned cDNAs and cloned gene sequences can also be used to detect and isolate homologous genes in the DNA of related species. Success will depend upon the divergence time of the two species and on the rate of evolution of coding sequences, which in turn is likely to be correlated with the clock rate of polypeptide sequence divergence. Thus rabbit adult β-globin cDNA is capable of detecting the entire family of human β-related globin genes (the ε-, $^{G}\gamma$-, $^{A}\gamma$-, δ- and β-globin genes, Barrie *et al.*, 1981), and has been used to isolate the human ε-globin gene from a recombinant λ phage library (Proudfoot and Baralle, 1979). Similarly, the chicken preproinsulin gene was isolated using a rat insulin cDNA probe (Perler *et al.*, 1980) and reptilian δ-crystallin DNA has been detected with a chicken probe (Williams and Piatigorsky, 1979). The most extreme cases of successful heterologous hybridization have been with cloned DNAs coding for highly conserved proteins. For example, sea urchin actin genes have been analysed using a *Drosophila* actin DNA probe (Durica *et al.*, 1980).

B Estimating DNA sequence divergence

If two related cloned fragments of DNA are to be compared for sequence divergence, the areas of homology must be located. This can be achieved by detailed restriction endonuclease mapping, cross-hybridization, or most speedily by electron microscopy of heteroduplexes formed between the two sequences. The latter method also gives a useful pictorial display of conserved and divergent sequences around related genes and has been used to compare, for example, the mouse β_d^{maj}- and β_d^{min}-globin genes (Leder *et al.*, 1980) and two chicken δ-crystallin genes (Jones *et al.*, 1980).

Once homologous sequences have been located, DNA sequence divergence is most accurately determined by total sequence analysis. The recent development of rapid DNA sequencing methods (Sanger *et al.*, 1978; Maxam and Gilbert, 1980) has led to an explosion of primary data suitable for evolutionary studies (Grantham *et al.*, 1980a, b, 1981). This situation is analogous to the appearance of polypeptide sequences during the 1960s, and will necessitate the development of central data banks capable of storing and updating complete sequences, and melding new partial sequences as they appear. Such banks are currently being assembled by the US National Institutes of Health (Bethesda) in co-operation with the European Molecular Biology Organization (Heidelberg), and also by Dr M. O. Dayhoff (Washington DC) and Prof. R. Grantham (Lyon) (see *Nature* 289, 112 (1980)).

Once two related DNA sequences are available, DNA sequence divergence can be calculated simply by aligning the sequences and scoring the proportion of nucleotides which differ between the two sequences. This procedure is most reliable when the two sequences are known to have diverged only by base substitution, as is generally the case when comparing two homologous protein coding sequences. Problems arise when deletions or insertions have occurred, as is frequently found in the evolution of extragenic DNA and intervening sequences. Optimal sequence alignment is commonly performed by constructing a matrix of nucleotide-by-nucleotide comparisons between the two sequences (Konkel *et al.*, 1979; see Fig. 1).

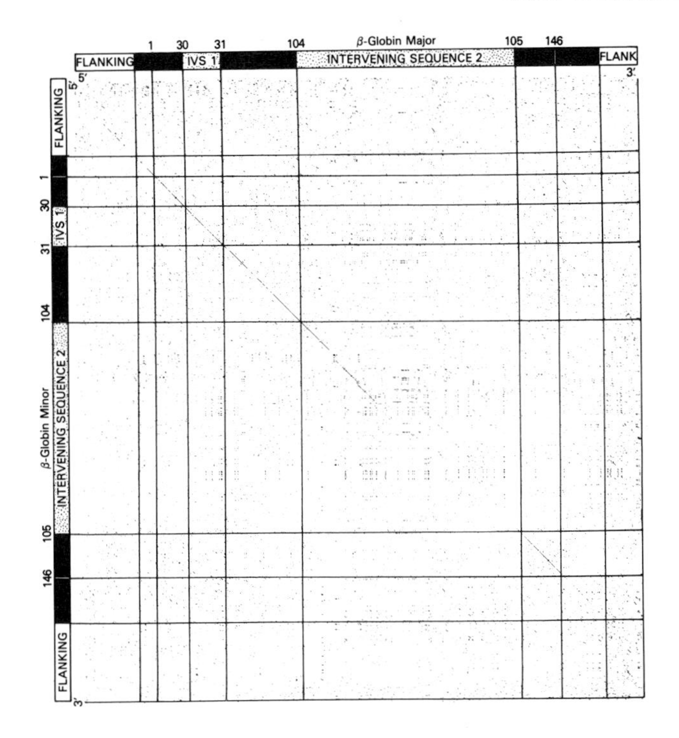

Figure 1 Dot matrix comparison of the DNA sequences of the mouse β_d^{maj}- and β_d^{min}-globin genes. The entire coding sequences, intervening sequences, and the immediate 5′ and 3′ flanking sequences are compared. Each dot represents the centre of a three-base identity between the two genes. Homology between these genes appears as a line at $-45°$ across the grid. Discontinuities in the line in the second intervening sequence indicate gene divergence by microdeletion/insertion, as well as by base substitution (see Fig. 7). From Konkel *et al.*, (1979); reprinted with permission of MIT press.

If the two sequences are identical, then a continuous line of identical bases appear on the diagonal, plus a scatter of spurious base identities over the entire matrix. Many of these unwanted identities can be removed by filtering out matches confined to an isolated base or pair of bases. If two sequences differ only by base substitutions, then the diagonal matrix line will show gaps corresponding to base differences. If a deletion or insertion has occurred, then the diagonal will show a discontinuity after which the line of homology will reappear parallel with the original line of identity. A measure of DNA sequence divergence can then be estimated by ignoring DNA tracts present in one sequence but not in the other. If several deletions/insertions have occurred, then a second independent estimate may be made of the frequency of these events per unit length of DNA. Other numerical methods of sequence alignment have been described by Van Ooyen *et al.* (1979).

Major problems arise when comparing two highly divergent sequences. Alignment of the two sequences, allowing for deletion/insertion differences, becomes subjective, and spurious sequence matches may well become incorporated within an exaggerated estimate of sequence homology. It is important to remember that two mutually random sequences containing equimolar amounts of A, T, G and C will show 25% sequence homology if randomly aligned, and $> 25\%$ homology if the alignment is optimized. A skewed base composition can also lead to an elevated level of spurious homology.

At present, DNA sequencing is only suitable for comparing sequences at most a few thousand base pairs in length. More extensive regions of DNA are more readily screened by comparing restriction endonuclease cleavage maps. These maps can be constructed either by analysing cloned DNAs, in which case every cleavage site present will be recorded, or by Southern blot analysis of total genomic DNA using a cloned DNA as a probe. In the latter case, only cleavage sites nearest the detected sequence will be located. The two related maps can be aligned to search for regions where sufficient site identities exist to establish that two homologous and identically arranged sequences are being compared. Restriction site differences within homologous regions can then be used to estimate DNA sequence divergence.

Upholt (1977) gave the first description for converting restriction endonuclease cleavage map differences to sequence divergence. More refined methods both for this conversion and for correcting divergence estimates for multiple substitutions (see below) have been developed by Nei and Li (1979). Consider two homologous sequences, x and y. n_x cleavage sites will be examined in x, and n_y in y, of which n_{xy} will be common to both. The proportion, S, of sites present in one map, which are also present in the other map, is given by:

$$S = \frac{2n_{xy}}{n_x + n_y}$$

If restriction endonucleases with hexanucleotide recognition sequences have been used, S can be related to the DNA sequence divergence, λ, by:

$$\lambda = 1 - S^{1/6}$$

(As pointed out by Nei and Li (1979), ambiguities in the original definition of S by Upholt (1977) have resulted in several erroneous λ values appearing in the literature.) This analysis can be extended to endonucleases that recognize sequences other than hexanucleotides, and to nearest-neighbour maps, derived by Southern blot analysis of total genomic DNA, in which not all restriction endonuclease cleavage sites are detected.

Restriction site differences are a sensitive indicator of DNA sequence divergence (Fig. 2) and are unsuitable for comparing DNA sequences that differ by more than 25%. If two sequences have diverged by microdeletion/insertion as well as base substitution, then the base substitution divergence, λ, will be overestimated. However, more substantial deletions and insertions should become obvious by comparing restriction endonuclease maps. Care also has to be taken when comparing Southern blot maps, if restriction endonucleases have been used which recognize sequences which can be modified in eukaryotic DNA (Bird and Southern, 1978; Van Der Ploeg and Flavell, 1980). Map differences might then reflect species-specific or tissue-specific differences in base modification rather than primary sequence.

DNA sequence divergences can also be estimated from restriction fragment differences, without ordering these fragments into physical maps of cleavage sites (Upholt, 1977; Nei and Li, 1979). Without mapping data, this approach is dangerous since many differences might be the result of major DNA rearrangements rather than base substitutions.

C Correcting DNA sequence divergence for multiple substitutions

If two initially identical DNA sequences have diverged by base substitution in two independent lineages, how many base substitutions

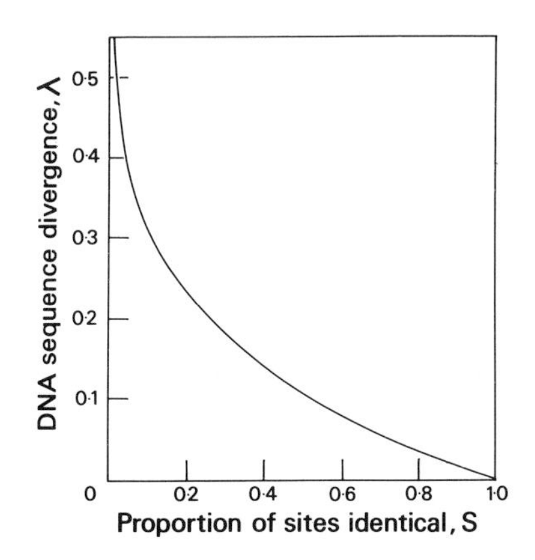

Figure 2 Relation between the proportion (S) of restriction endonuclease cleavage sites, present in one cleavage map, which are also present in a second map, and the DNA sequence divergence (λ) between the two maps, calculated according to the equation $\lambda = 1 - S^{1/6}$ (Upholt, 1977). This relation is only valid for restriction endonucleases which have hexanucleotide recognition sequences.

must have occurred to give the contemporary divergence of nucleotide sequence? A "random substitution" model is generally used (Kimura, 1977; Nei and Li, 1979) and makes two key assumptions: first, substitutions occur at random positions in a DNA sequence; second, a given substitution can result in one of three equally probable changes — for example, an A residue can be replaced at random by C, G or T. Consider two originally identical sequences L nucleotides long which have undergone n substitutions leaving σ_n nucleotides still identical between the two sequences. The $(n + 1)$th substitution could either occur in one of the σ_n identical sites, with a probability of σ_n/L and decreasing σ_n by 1, or in one of the non-identical sites, with a probability of $1 - (\sigma_n/L)$. Two-thirds of the latter substitutions will maintain the sequence difference, whereas one-third will cause the originally non-identical sites to become identical. Thus the $(n + 1)$th substitution will give a decreased σ according to:

$$\sigma_{n+1} = \sigma_n - \frac{\sigma_n}{L} + \frac{1}{3}\left(1 - \frac{\sigma_n}{L}\right)$$

The equation has the following approximate solution:

$$1 - \frac{\sigma_n}{L} = \frac{3}{4}\left(1 - e^{-4n/3L}\right)$$

valid for large values of L. Substituting the DNA sequence divergence, λ, for $1 - (\sigma_n/L)$, and the number of substitutions per nucleotide site, K, for n/L, gives:

$$K = -\frac{3}{4}\ln\left(1 - \frac{4}{3}\lambda\right)$$

A plot of K versus λ is given in Fig. 3. As the number of substitutions increases, λ becomes an increasingly greater underestimate of K as a result of multiple substitutions, and after an infinite number of substitutions λ tends to 0.75, as expected for two mutually random sequences.

This "random substitution" model will be inaccurate particularly at high levels of substitution, since the three possible base changes per substitution (one transition and two transversions) do not occur with equal probability. Van Ooyen *et al.* (1979) have compared the rabbit and mouse adult β-globin genes, and show that 47% of changes were transitions, in contrast to the "random substitution" model prediction of $33\frac{1}{3}$%. Substitution biases have also been noted by Jukes (1980), and will tend to lead to an apparent reduction in the rate of sequence divergence. For example, if the transitions A $\rightleftharpoons$ G are more likely than the transversions A or G $\rightleftharpoons$ T or C, then an A residue initially common to two sequences will preferentially change

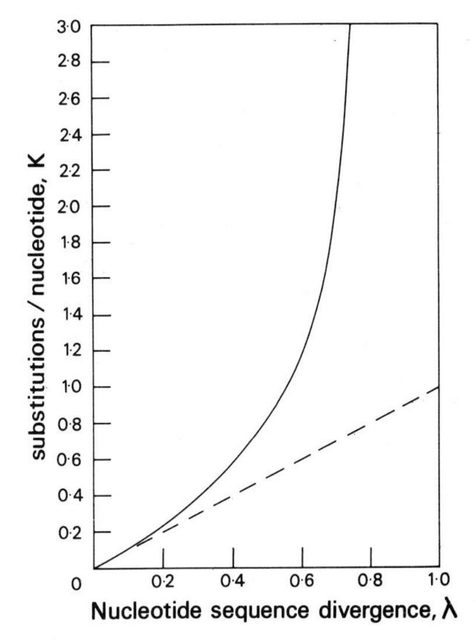

Figure 3 Relation between nucleotide sequence divergence (λ) and the incidence of nucleotide substitutions (K). The dashed line shows the relation between λ and K uncorrected for multiple substitutions. The continuous line is corrected for multiple substitutions, assuming a "random substitution" model (see text), according to the equation $K = -\frac{3}{4} \ln (1 - \frac{4}{3} \lambda)$ (see Kimura, 1977).

to G in one sequence at the first substitution. Similarly, a second substitution at this site will preferentially revert the G to A or alter the second sequence to G. In either case, divergence has occurred less rapidly than predicted by the "random substitution" model. Thus the number of replacements per site, K, deduced from the DNA sequence divergence, λ, will be an underestimate.

III Evolution of protein coding sequences

A Replacement site and silent site substitutions

Although many proteins are highly conserved in evolution, redundancy of the genetic code can in principle allow the coding sequences for identical proteins to differ substantially. This phenomenon was first established in a comparison of histone H4 mRNA sequences coding for the highly conserved histone H4 and isolated from two sea urchins, *Lytechinus pictus* and *Strongylocentrotus purpuratus* (Grunstein *et al.*, 1976). These two species diverged relatively recently (about 60 million years ago) yet showed an 11.5% divergence of H4 mRNA sequences. However, none of the substitutions resulted in an amino acid replacement, and instead were all silent, causing synonymous codon changes.

Theoretically there are 526 possible single-base substitutions in the 61 sense codons, 134 (25%) of which are silent (Jukes and King, 1979). If histone H4 is conserved in evolution solely due to an inability of the genes to mutate, then those mRNA sequence differences that do exist should be distributed as above, with most substitutions causing amino acid replacements. Clearly, the histone H4 gene is capable of evolving, but selection has rigorously eliminated all base substitutions which have resulted in altered codon specificity. The main questions now are whether the silent changes have become fixed by neutral drift, and whether all possible silent changes are also neutral.

A similar phenomenon was found in a comparison of the human and rabbit β-globin mRNA sequences (Kafatos *et al.*, 1977). Of 48 base substitutions in the coding sequence, 32 (67%) were silent. This deficit of replacement substitutions again points to selection acting against the majority of amino acid substitutions in globin. In addition, the silent nucleotide substitutions are not scattered randomly over the coding sequence, but are clustered, with replacement-free segments being relatively deficient in silent substitutions. Also, the overall extent of silent site divergence between human and rabbit β-globin coding sequences is less than that predicted by the rate of divergence of hypervariable residues in fibrinopeptides, which are thought to change largely by neutral evolution. Thus some silent substitutions appear to have been eliminated by selection, perhaps as a result of interference with mRNA or precursor mRNA structure, or RNA processing and export. Alternatively, a silent substitution might generate a synonymous codon for which no abundant tRNA is available. Most species examined to date show a marked bias in codon utilization patterns, and although the pattern for one species tends to reflect tRNA availabilities and is probably shared by the majority of genes in that species, this pattern can shift from species to species (Grantham *et al.*, 1980a, b, 1981). If codon utilization patterns change little in evolution, then selection might tend to eliminate seldom used codons. However, should the pool of available tRNAs drift during evolution, this would apply a positive selective pressure for certain silent site substitutions. Thus, while the majority of silent substitutions which have become fixed in evolution are likely to be neutral, as argued by Jukes and King (1979), a proportion might well be adaptations to shifting patterns of codon utilization, or conceivably to physiological changes in mRNAs and their precursors, rather than in protein. Conversely, some amino acid residues in proteins might be highly conserved in evolution, not because they are essential to protein function, but because the respective coding sequences have some vital function at the DNA or RNA level.

This preferential accumulation of silent nucleotide changes in protein coding sequences has been confirmed in a survey of a wide variety of prokaryotic, eukaryotic and viral genes (Jukes, 1980).

B Rates of replacement site substitution

Amino acid replacement site substitutions accumulate more slowly in evolution than do silent site substitutions. Perler *et al.* (1980) have compared coding sequences for various mammalian and avian α- and β-globin genes and preproinsulin genes. Replacement site and silent site divergences were calculated and corrected for multiple substitutions using the "random substitution" model (see above). Plots of sequence divergence against the estimated divergence times of the genes being compared are shown in Fig. 4. Replacement site substitutions appear to accumulate linearly with time in coding sequences for the four polypeptides examined (insulin A, B and C peptides and globins; a plot of globin substitution against time has also been presented by Efstratiadis *et al.*, 1980). The substitution

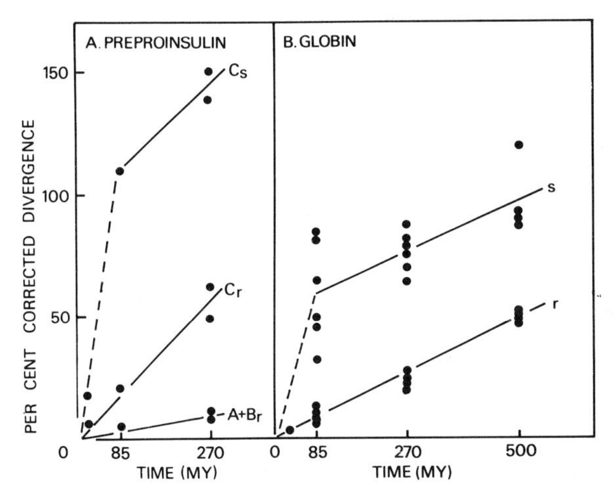

Figure 4 Rates of DNA sequence divergence during the evolution of insulin and globin genes. Silent site (s) and replacement site (r) divergences are corrected for multiple substitutions using the "random substitution" model of Kimura (1977), and plotted against the estimated divergence times in millions of years (MY). Plots are shown for silent and replacement changes in the rapidly evolving C peptide of preproinsulin, for replacement changes in the slowly evolving prepro-insulin A and B peptides, and for silent and replacement changes in globins. Major time ordinates are 85 MY (divergence of the mammals, Romero-Herrera *et al.*, 1973), 270 MY (divergence of birds and mammals, Wilson *et al.*, 1977) and 500 MY (divergence of α- and β-globin genes, see Fig. 5). More recent divergences are for the recently duplicated rat preproinsulin I and II genes, and for the mouse β_d^{maj}- and β_d^{min}-globin genes. From Perler *et al.* (1980), reprinted with permission of the MIT press.

rate is characteristic for each polypeptide and is roughly proportional to the rate of amino acid sequence divergence derived from protein studies (Wilson *et al.*, 1977). These observations are, of course, expected in view of the protein evolutionary clock hypothesis plus the fact that most of the divergence times used by Perler *et al.* (1980) were estimated from protein sequence data. Efstratiadis *et al.* (1980) have used the replacement substitution clock to construct an evolutionary tree for the human α- and β-globin gene family (Fig. 5). As expected, this phylogenetic tree corresponds closely to the phylogeny

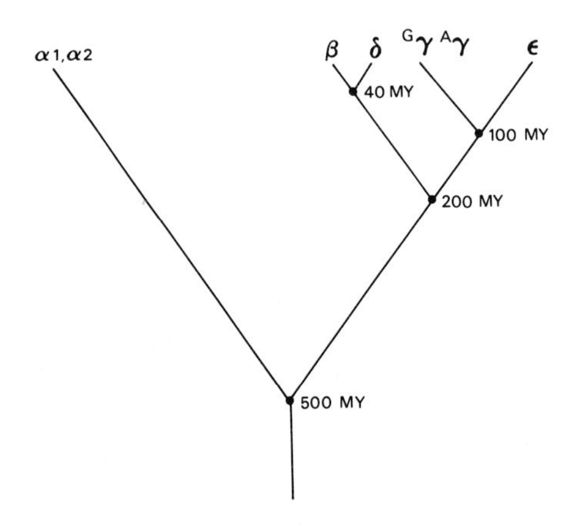

Figure 5 An evolutionary tree of the human globin genes, derived by a comparison of corrected replacement site divergences of individual pairs of genes. The branch points represent the times in millions of years (MY) before the present, either when gene duplications occurred or when duplicated gene pairs ceased to evolve in concert. From Efstratiadis *et al.* (1980), reprinted with permission of the MIT press.

established by amino acid sequence comparisons. These family trees are only accurate if the divergence rates for all of these related genes are equal. This has not been established, for example, for the ϵ-globin gene, which has only been sequenced in man (Baralle *et al.*, 1980) and might be evolving at a rate atypical of adult α- and β-globin genes.

C Rates of silent site substitution

As shown in Fig. 4, silent site substitutions tend to accumulate rapidly in evolution in an apparently non-linear fashion. In both preproinsulin and globin coding sequences, the degree of divergence becomes very substantial after about 100 million years. From this point, correction for multiple substitutions becomes increasingly error-prone as the measured sequence divergence approaches asymptotically to 75% (Fig. 3). These errors will be compounded from statistical errors in divergence estimates (not given in the plots of Perler *et al.*, 1980, or Efstratiadis *et al.*, 1980) and from departures from the "random substitution" model of sequence divergence (see above).

What is the initial rate of silent site substitution? For the insulin C peptide the most accurate divergence estimate comes from a comparison of the closely related rat preproinsulin I and II genes, which show a corrected silent site divergence of 18% compared with a 3.2% replacement site divergence. Knowing the replacement clock rate for the C peptide gives a divergence time for the rat I and II genes of about 15 million years, corresponding to the time either of gene duplication or the last round of gene correction between these loci (see below). The initial rate of silent site divergence is therefore about 1% in 0.8 million years, giving a Unit Evolutionary Period (UEP, as defined by Wilson *et al.*, 1977) of 0.8. A similar initial UEP of 1.3 (1% divergence in 1.3 million years) has been reported for globin genes, based primarily upon a comparison of silent site substitutions in human β- and δ-globin genes (Efstratiadis *et al.*, 1980).

These estimates of the initial rates of silent site substitutions are very approximate. In the comparison of rat preproinsulin I and II genes, the C peptide coding region examined is only 84 bases long (Perler *et al.*, 1980), giving a very approximate estimate of replacement site divergence and therefore divergence time (Lomedico *et al.*, 1979). In the comparison of human δ- and β-globin genes (Efstratiadis *et al.*, 1980), the replacement site and silent site divergences differ markedly for the three exons in these genes: 5.5%, 2.2% and 5.1% for replacement site divergences and 16%, 16% and 112% for silent site divergences in the three exons respectively.

Clearly, many more gene sequences must be examined before an accurate initial rate of silent site substitution can be derived. Many questions will need answering. Is the rate constant for all genes? Can it act as an evolutionary clock for the phylogenetic analysis of closely related species? Is the clock geared to absolute time or generation time? In this last context, it is worth noting that the apparently similar insulin and globin rates were established from rodents and primates respectively, the generation times of which differ by an order of magnitude.

D Neutral versus adaptive substitution

The current best estimate for the initial rate of silent site substitution gives a UEP of about 1.0. This figure is likely to be an inaccurate estimate of the maximal rate of neutral evolution of DNA sequences devoid of any function. First, many possible silent site substitutions might be deleterious and be eliminated by selection, as noted above. Second, some silent substitutions are likely to be selectively advantageous; even a minute advantage particularly in large populations can improve substantially the chance of fixation (see Gale, 1980). A more accurate estimate of the rate of neutral drift might come from a phylogenetic study of the possibly non-functional pseudogenes (see below).

This estimate of neutral drift rate is important since the neutral theory shows that the rate of fixation of neutral variants per generation in a given lineage is equal to the mutation rate per gamete (Kimura, 1969; King and Jukes, 1969). The current rough estimate of UEP = 1.0 gives a mutation rate of 5×10^{-9}/nucleotide/year or 10^{-7}/nucleotide/gamete (generation) in man. For an average structural locus about 1000 base pairs long, the mutation rate to a variant protein electromorph (Neel *et al.*, 1980) will be about 2×10^{-5}/locus/gamete. This estimate is not too dissimilar from the current upper limit of 0.6×10^{-5}/locus/gamete derived from extensive screenings of human populations for new electrophoretic protein variants (Neel *et al.*, 1980). This limit may in turn be an underestimate since not all electrophoretic variants would necessarily have been detected by a single round of electrophoretic screening. Whether any real difference exists between these two independent estimates of mutation rate in man is not known.

The accumulation of silent site substitutions apparently slows down after about 100 million years of divergence, and thereafter proceeds at about the same rate as replacement site substitutions (Perler *et al.*, 1980; Efstratiadis *et al.*, 1980; see Fig. 4). Part of this apparent decrease might be due to an underestimated correction for multiple substitutions, resulting from the limitations of the "random substitution" model (see above). If the decrease is genuine, then the simplest explanation is that there are (at least) two classes of silent site substitution. One occurs exceedingly rapidly, with a UEP as low as 0.6 for globin, and may represent neutral sites which become randomized within 100 million years. The second set of silent sites change more slowly (UEP = 10) and might reflect the gradual appearance of adaptive changes. Does the apparent similarity between this slow rate and the rate of replacement site substitution indicate that the replacement clock is not driven by neutral drift? Perler *et al.* (1980) argue for selection as follows: replacement site and silent site substitutions occur in globin genes with UEPs of about 1 and 10 respectively. This suggests that at least 90% of possible replacement changes are eliminated by selection, and thus that no more than 10% of amino acid replacements are neutral. However, amino acid substitutions do not saturate at 10% during evolution but continue accumulating linearly with time (Dayhoff, 1972; Wilson *et al.*, 1977). Therefore, non-neutral replacements do seem to occur, suggesting that the replacement clock is not driven primarily by neutral drift. An objection to this argument is that neutral changes could open up new neutral sites (including the slowly evolving silent sites) at which changes were originally disadvantageous.

Over short time scales, localized fluctuations in evolutionary pressures can give rise to unexpected patterns of base substitution. For example, two common alleles of rabbit adult β-globin exist which differ from each other by four amino acid residues. Both alleles have been cloned and sequenced (Efstratiadis *et al.*, 1977; Hardison *et al.*, 1979; Van Ooyen *et al.*, 1979). The allelic coding sequences differed by four single replacement substitutions, as expected, but surprisingly, there were no silent changes. It seems that these replacement substitutions have attained a polymorphic frequency very recently in the evolution of the rabbit, before silent changes had the opportunity to appear, and as the result of positive selection for this combination of variants. Bodmer (1981) has pointed out that mechanisms such as positive selection and asymmetric gene conversion in heterozygotes during meiosis can lead to very rapid sequence evolution, and has cautioned against equating very rapidly evolving sequences with functionless DNA evolving by neutral drift (see Gale, 1980).

IV Evolution of intervening sequences

A Occurrence of intervening sequences

Since the discovery in 1977 of split genes coding for proteins, many examples of discontinuous genes have been reported (see Abelson, 1979; Crick, 1979; Breathnach and Chambon, 1981). To date, they have been found in genes isolated from vertebrates, insects, echinoderms and fungi, and most recently in plants (Sun *et al.*, 1981), as well as in organelle DNA and viral DNA. The only higher eukaryotic genes which definitely do not contain intervening sequences are the histone genes (Kedes, 1979) and a human leukocyte interferon gene (Nagata *et al.*, 1980). Most mouse brain poly A^+ RNAs seem to be encoded by split genes, and the mouse genome probably contains at least tens of thousands of discontinuous genes (Maxwell *et al.*, 1980). In contrast, no example of a discontinuous gene in a prokaryote has yet been reported. Assuming that intervening sequences did not appear totally independently in a wide variety of different eukaryotic lineages, this strongly suggests that split genes can be traced at least as far back as the earliest eukaryotes, in existence perhaps as much as 3000 million years ago (Woese and Fox, 1977). The apparent lack of split genes in prokaryotes can then be interpreted either as the appearance of split genes concommitant with the emergence of early eukaryotes, or as a subsequent elimination of discontinuous genes specifically in prokaryotes (Doolittle, 1978). Evidence does exist which indicates that intervening sequences can occasionally be lost during evolution (see below).

B Evolutionary stability of gene structure

There are several examples of families of split genes which have evolved by gene duplication and where all members still retain a common and presumably ancestral pattern of intervening sequences. Globin genes are the best studied family. All active vertebrate α- and β-globin genes examined in species ranging from *Xenopus laevis* to man contain two intervening sequences within the protein coding sequence (Table 1). In every case, the intervening sequences occur at precisely homologous positions within these genes. As noted by Leder *et al.* (1978) this establishes that the ancestral globin gene, which duplicated to give α- and β-globin genes about 500 million years ago (Efstratiadis *et al.*, 1980), must also have been interrupted by two intervening sequences. Thus, over a period of 2600 million years of evolution (the sum of the times of independent divergence of all genes listed in Table 1), no intervening sequences have been gained or lost by active globin genes.

Structural stability is also suggested by other gene comparisons. For example, two non-allelic vitellogenin genes have been cloned from *Xenopus laevis*; each contains at least 33 intervening sequences, at apparently identical positions within these genes (Wahli *et al.*, 1980). Similarly, two chicken δ-crystallin genes each contain 14—15 intervening sequences probably at homologous positions within the two genes (Bhat *et al.*, 1980; Jones *et al.*, 1980). The chicken ovalbumin gene contains seven intervening sequences (Gannon *et al.*, 1979), and this structure has also been preserved in the closely linked X and Y genes, whose function is unknown (LeMeur *et al.*, 1981) and which are apparently the products of a series of duplications in the ovalbumin gene region (Royal *et al.*, 1979; Heilig *et al.*, 1980). The X and ovalbumin genes are estimated to have diverged 55 million years ago.

C Loss of intervening sequences

Two clear cases have been described where one or more intervening sequences have been cleanly removed from a gene during evolution. Lomedico *et al.* (1979) have compared the structures of the rat

Table 1 Sizes of intervening sequences in vertebrate globin genes and pseudogenes.*

Gene		Size (bp) IVS 1	IVS 2	References
β-related				
man:	β	130	850	Lawn *et al.* (1980)
	δ	128	889	Spritz *et al.* (1980)
	${}^{A}\gamma$	122	866 (876)†	Slightom *et al.* (1980)
	${}^{G}\gamma$	122	886 (904)†	
	ϵ	122	850	Baralle *et al.* (1980)
rabbit:	$\beta 1$	126	573	Van Ooyen *et al.* (1979)
	$\psi\beta 2$	100	~ 770	Lacy and Maniatis (1980)
mouse:	β_{d}^{maj}	116	653	Konkel *et al.* (1979)
	β_{d}^{min}	116	628	
goat:	$\psi\beta^{x}$	125	~ 650	Cleary *et al.* (1980)
chicken:	β	~ 100	~ 800	Dodgson *et al.* (1979)
X. laevis:	β^{1}	~ 190	~ 930	Patient *et al.* (1980)
α-related				
man:	$\alpha 2$	117	140	Liebhaber *et al.* (1980)
	$\psi\alpha 1$	127	134	Proudfoot and Maniatis (1980)
mouse:	α	122	134	Nishioka and Leder (1979)
	$\psi\alpha$	0	0	Nishioka *et al.* (1980); Vanin *et al.* (1980)
X. laevis:	α^{1}	169	339	R. M. Kay, R. K. Patient, J. G. Williams and R. Harris, unpublished data

*IVS 1 and IVS 2 are present at homologous positions in all globin genes studied. Genes prefixed by ψ are pseudogenes. Mouse $\psi\alpha$ contains no intervening sequences.

†The two sizes of IVS 2 for the human γ-globin genes represent the sizes of allelic variants.

preproinsulin I and II genes, which are estimated to have duplicated about 15 million years ago. Both genes contain a 119 bp intervening sequence in the 5′ non-translated mRNA region. In addition, the preproinsulin II gene, but not the I gene, contains a 499 bp intervening sequence within the coding sequence. Since the chicken and human preproinsulin genes are interrupted twice (Bell *et al.*, 1980; Perler *et al.*, 1980), the rat preproinsulin I gene must have lost its second intervening sequence recently, after the duplication of the rat I and II genes.

An even more extraordinary case of precise removal of intervening sequences has been found in a mouse α-globin-related pseudogene (Nishioka *et al.*, 1980; Vanin *et al.*, 1980). This gene is 84% homologous to the mouse adult α-globin gene, and is clearly a relatively recent product of α-globin gene duplication. However, it has precisely lost both intervening sequences present in all active globin genes so far examined (Table 1). In addition, a number of deletions, insertions and frameshifts have accumulated to render this α-globin gene sequence non-functional. It is not known whether the α-globin gene was silenced before the intervening sequences were removed, or whether the removal of these sequences was responsible for the inactivation.

The structure of actin genes has also varied during evolution. The yeast actin gene contains only one intervening sequence, interrupting the fourth codon (Gallwitz and Sures, 1980; Ng and Abelson, 1980). A sea urchin actin gene contains at least one intervening sequence, but this interrupts the 121st codon (Durica *et al.*, 1980). A *Drosophila* actin gene also contains at least one intron at an unknown position towards the 5′ end of the gene (Fyrberg *et al.*, 1980). In contrast, a chicken α-actin gene contains at least three very short introns (Ordahl *et al.*, 1980). Whether these changes have involved gain or loss of introns during evolution is not known.

How are intervening sequences lost so precisely during evolution? Several mechanisms have been suggested by Nishioka *et al.* (1980), Vanin *et al.* (1980) and Leder *et al.* (1980). Any mechanism must have occurred in the germ line, and presumably involved, directly or indirectly, the splicing enzyme normally used for excising transcribed intervening sequences from RNA. Perhaps low levels of many mRNAs are produced in germ line cells. These could be reverse transcribed to give a double stranded copy DNA which could recombine into an homologous split gene and replace the coding regions containing the intervening sequences. Alternatively, during germ line DNA replication, the mature mRNA might anneal to the corresponding coding strand at the replication fork: the intervening sequences could then be repaired out of the duplex to give a continuous gene exactly co-linear with mRNA (Leder *et al.*, 1980; see Fig. 6). Yet another possibility is that the splicing enzyme itself could operate directly on the non-transcribed strand of the gene during replication. Clearly, genes which have been processed to remove intervening sequences are seldom fixed in evolution, either because the repair mechanism operates rarely, or because the bulk of genes so modified are at a selective disadvantage compared with their split relatives.

There is still considerable confusion over the role if any of intervening sequences in gene function. Some genes, for example those for interferon, function without these sequences. The second intervening sequence in the rat preproinsulin I gene has been lost without obvious gene dysfunction. However, the presence of an intervening sequence in hybrids between SV40 and the mouse β-globin gene seems essential in some way for the correct processing

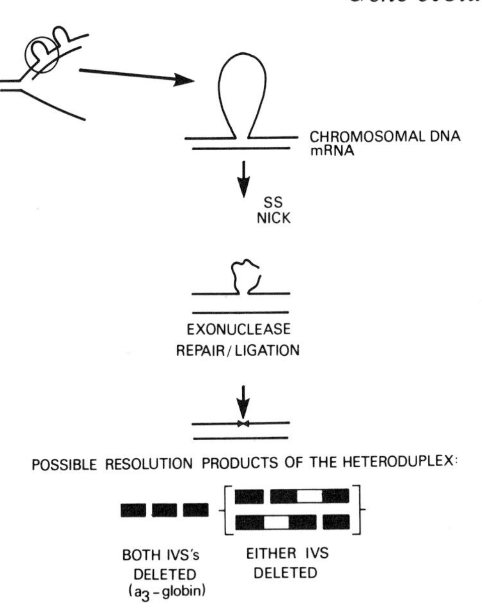

Figure 6 A model for the excision of intervening sequences from a split gene during replication. A replication fork passes over a gene containing two intervening sequences. The coding strand hybridizes with its corresponding mature mRNA and the intervening sequences are displaced as single stranded DNA loops. Excision and ligation of a loop results in the precise elimination of an intervening sequence from the gene. Either both intervening sequences are removed, as in the mouse $\psi\alpha$ gene, or only one is eliminated, as in the rat preproinsulin I gene. From Leder *et al.* (1980), copyright 1980 by the American Association for the Advancement of Science.

and/or export of transcripts from this gene in monkey cells (Hamer and Leder, 1979a, b). In contrast, the single 14 bp long intervening sequence within the yeast suppressor $tRNA^{Tyr}$ gene can be removed without destroying suppressor function (Wallace *et al.*, 1980). There are several instances in animal viruses where an intervening sequence for one mRNA forms part of a second mRNA sequence (see Abelson, 1979). Similarly, an intron within the cytochrome b gene in yeast mitochondrial DNA codes for part of a "mRNA maturase" involved in splicing cytochrome b mRNA (Lazowska *et al.*, 1980). There are also examples of nuclear introns that carry coding sequences. The *C*-terminal region of the μ chain in the secreted form of IgM is coded by part of an intervening sequence in the gene specifying membrane-bound μ; the two forms of μ chain mRNA are probably generated by alternative RNA splicing pathways from a common transcript (Early *et al.*, 1980; Rogers *et al.*, 1980). Alternative processing pathways might also account for the differences between the 5′ regions of mouse liver and salivary α-amylase mRNAs (Hagenbüchle *et al.*, 1981).

D The curious case of leghaemoglobin

Leghaemoglobin (Lb) is formed in the nitrogen-fixing root nodules of legumes and is encoded by the plant genome (Sidloi *et al.*, 1978). Although Lb is structurally and functionally similar to animal myoglobin and haemoglobin (Appleby, 1974), it has often been assumed that this similarity was the product of convergent evolution. However, a Lb gene has been isolated from soybean DNA (Sullivan *et al.*, 1981) and has been shown to contain three intervening sequences, two of which correspond in position to the animal haemoglobin intervening sequences (Marcker, 1980). Either the gene structure, including intron position, as well as protein structure have shown convergence, or the animal and plant genes are in some way directly related. It is difficult to see how this relationship can be traced back to the common ancestor of plants and animals since the vast majority of plants do not possess Lb. Instead, it is possible that horizontal gene transfer has occurred between animals and plants; for example perhaps insect globin sequences (Dayhoff, 1972) were recently transferred to a legume via an insect-borne plant pathogenic virus such as a rhabdovirus (Francki and Randles, 1980). If so, then insect globin genes might also have three intervening sequences, not two, and perhaps one of these sequences was eliminated in the lineage leading to the vertebrates at some time before the αβ-globin gene duplication.

E Divergence of intervening sequences

Comparisons of intervening sequences in a number of homologous gene pairs have shown how introns diverge in evolution. Intervening sequences in the rabbit and mouse β-globin genes are highly divergent compared with the coding sequences (Van Den Berg *et al.*, 1978; Van Ooyen *et al.*, 1979), although the sequences at splice junctions (Breathnach and Chambon, 1981) tend to be conserved. Attempts to align the equivalent rabbit and mouse sequences suggested that introns had diverged not only by base substitution but also by small deletions and insertions, reminiscent of those accumulated in the mouse α-globin pseudogene (Nishioka *et al.*, 1980; Vanin *et al.*, 1980). These multiple deletions and insertions make it difficult to estimate reliably a rate of DNA divergence by base substitution, although it is clear that these intervening sequences are diverging much faster than replacement sites in coding sequences.

Similar conclusions were drawn from a comparison of the closely related mouse β_d^{maj}- and β_d^{min}-globin genes, which are thought to have duplicated about 50 million years ago (Konkel *et al.*, 1979). Alignment of the two gene sequences (Fig. 7) showed that the second intron had diverged substantially, again by substitution and microdeletion/insertion. In contrast, the first intervening sequence showed little divergence, initially suggesting that this sequence has been functionally constrained in evolution. With the discovery of concerted evolution (see below), it now seems more likely that the 5′ ends of the two β-globin genes, including the first intervening sequence, have recently been corrected against each other, removing evidence of divergence.

Correlated with the high rate of intron evolution is what appears to be a preferential accumulation of genetic variation within intervening sequences, revealed as restriction endonuclease cleavage site polymorphisms. Such variants have been reported in the ovalbumin gene (Lai *et al.*, 1979) and in the β-related globin genes of man and other primates (Jeffreys, 1979; Barrie *et al.*, 1981).

Despite the high frequency of microdeletions/insertions in globin genes during evolution, the lengths of intervening sequences remain surprisingly constant (Table 1). The first intervening sequence is about 116—130 bp long in all mammalian α- and β-globin genes studied to date, despite the extreme age of the αβ-globin gene duplication. The second globin intron also varies little in length within the α- or β-globin gene families, although the α-sequence is consistently shorter than the β. Van Den Berg *et al.* (1978) have suggested that intron length, rather than sequence, might be important in some way to globin gene function. However, substantial changes in intron lengths during evolution have been noted in preproinsulin genes (Perler *et al.*, 1980), vitellogenin genes (Wahli *et al.*, 1980), δ-crystallin genes (Jones *et al.*, 1980) and the related prolactin and growth hormone genes in the rat (Chien and Thompson, 1980).

Efstratiadis *et al.* (1980) have noted that deleted sequences within globin intervening sequences, coding sequences and flanking sequences are frequently bounded by a short direct repeat from 2 to 8 bp. They suggest that microdeletions might be generated by slipped mis-pairing of these sequences during DNA replication (Fig. 8).

F Evolution of split genes

1 *Exon shuffling*

The prevalence of split genes within eukaryotic genomes has led to considerable speculation about the manner in which these genes have evolved. Gilbert (1978) suggested that mutations creating new RNA

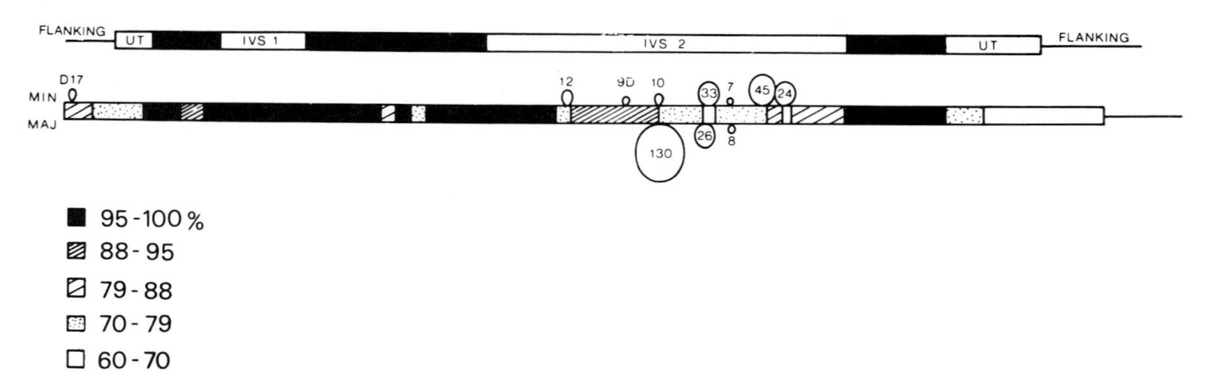

Figure 7 Homology map of the mouse β_d^{maj}- and β_d^{min}-globin genes. The structure of these genes is shown in the upper figure, and includes coding sequences (solid boxes), intervening sequences (IVS), and the 5′ and 3′ untranslated sequences (UT) in β-globin mRNA. The lower figure shows percentage DNA sequence homology across the gene. Sequences present in one gene but not the other are shown as loops. This homology map was derived from the dot matrix comparison shown in Fig. 1. From Konkel *et al.*, (1979), reprinted with permission of MIT press.

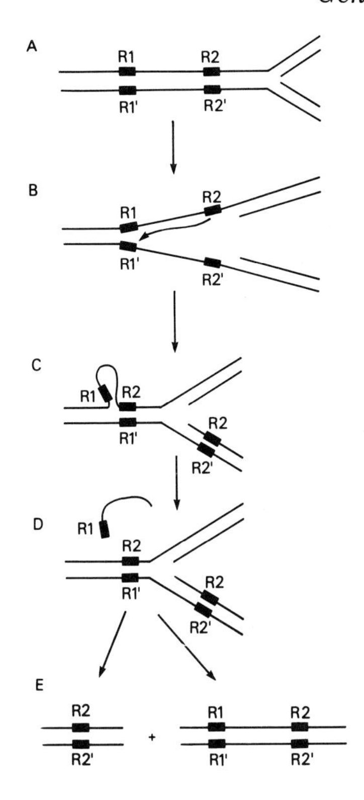

Figure 8 The "slipped mispairing" model for generating small deletions during DNA replication. DNA containing two short direct repeats (R1 and R2) and their complements (R1′ and R2′) is replicated (A). As the replication fork reaches R1, R2 mispairs with R1′ (B). The displaced DNA loop (C) is excised (D) and subsequent replication generates one daughter molecule in which the region between R1 and R2 is deleted (E). From Efstratiadis *et al.*, (1980), reprinted with permission of the MIT press.

splicing pathways in transcriptional units could lead to new combinations of coding sequence blocks (exons) appearing within mature mRNAs. As a consequence, novel polypeptides could be evolved rapidly, without necessarily destroying the original gene function. Similarly, recombination between introns could generate new combinations of exons again specifying novel genetic functions. Split genes are therefore seen as the consequence of storing protein coding information, not in genes, but in exons which can be shuffled at the DNA and RNA level to generate biochemical diversity during evolution (Darnell, 1978). As a result, new genes can be created more rapidly than is possible by the slow processes of gene duplication and sequence divergence (Reanney, 1979). Doolittle (1978) argues that evolution by exon shuffling predates the divergence of prokaryotes and eukaryotes perhaps 2500—3000 million years ago, and originally emerged as a strategy for ensuring that useful exons were preserved by reiteration, and at least occasionally expressed correctly, in primitive organisms in which information storage and expression were less faithfully executed than in contemporary organisms.

Blake (1979) proposed that exons might encode discrete stable domains within proteins; exon shuffling during evolution would thereby create, not haphazard arrays of peptide sequences, but a collection of discrete domains more likely to be assembled into a stable functional protein. There are several instances where exons do seem to code for discrete functional or structural regions within proteins. For example, the central exon in globin genes codes for the haem-binding domain of globin (Craik *et al.*, 1980), and the distribution of intersubunit contacts in haemoglobin appears to be distributed non-randomly with respect to the three exons (Eaton, 1980). Similarly, the four exons of the chicken lysozyme gene each appear to specify different functional regions of the protein. Exon 1 codes for the signal peptide of prelysozyme and the amino-terminal region of lysozyme. Exon 2 codes for the active site plus substrate binding residues. Residues conferring additional substrate specificity are coded by exon 3, and the carboxy-terminal region of lysozyme by exon 4. A strong correlation between exons and domains is also seen in immunoglobulin genes. A single intron in V genes separates coding sequence for the signal peptide from the major portion of the V region (Tonegawa *et al.*, 1978). Similarly, each of the three heavy chain C_γ domains, plus the hinge region, are encoded by separate exons (Sakano *et al.*, 1979b; Early *et al.*, 1979). To some extent, the correlations between exons and domains in C_H genes might be the result of tandem duplications of a single C exon during evolution, rather than the C_H gene having been assembled by the reshuffling of a series of unrelated exons (Sakano *et al.*, 1979a).

The highly-split chicken ovalbumin gene also appears to have evolved by duplication and fusion of an already-split gene (Cochet *et al.*, 1979), as does the chicken ovomucoid gene, in which coding sequences for the three trypsin-binding domains each contain one intervening sequence and are also separated from each other and from a leader sequence by additional intervening sequences (Stein *et al.*, 1980). It is entirely likely that not only exon shuffling, but also duplication—fusion, have been responsible for the complex internal organization of many contemporary genes such as the very highly split α-collagen gene (Schafer *et al.*, 1980; Vogeli *et al.*, 1980). In contrast, the human POMC gene codes for the pro-opiomelanocortin peptide which is cleaved to give initially corticotropin and β-lipotropin and subsequently to give a range of melanotropins, lipotropins and endorphin: despite the clear functional domains within the POMC peptide, no intervening sequences have been found within the POMC coding sequence (Chang *et al*., 1980).

2 *Insertion sequences and intervening sequences*

Darnell (1978) and Crick (1979) have suggested that intervening sequences might have arisen by the insertion of transposable elements into originally continuous genes. Pre-existing RNA splicing systems would have rendered such events non-lethal. Subsequent evolution would then have made the insertion elements non-transposable but would not necessarily have eliminated them entirely from genes. This model readily accounts for the strict co-linearity of gene and protein in all split genes examined to date, even though it is at least theoretically possible for RNA splicing to rearrange the order of exons in the final mature mRNA. The exon shuffling model can only account for this co-linearity by assuming that the splicing mechanisms can remove RNA only from between *adjacent* active splice points within transcripts (see Lewin, 1980). However, there is no evidence for intervening sequences being acquired by nuclear genes during evolution (see above), and instead it seems that the number of introns in a gene decreases, not increases, during evolution.

It is possible that some of the introns in the cytochrome b gene in yeast mitochondrial DNA might have arisen by insertion. Three of the five introns in this gene, including the mRNA maturase sequence, are absent in some yeast strains (Nobrega and Tzagoloff, 1980). Borst and Grivell (1981) conjecture that these three dispensable introns were once transposons, and that perhaps the mRNA maturase has evolved from a transposase originally used to excise DNA, not RNA.

If intervening sequences in nuclear genes were once transposable elements, then they might be regarded as "junk" or "selfish" DNA (Doolittle and Sapienza, 1980; Orgel and Crick, 1980) which at some time in the past was capable of spreading throughout the genome without necessarily altering the phenotype. This notion of "junk" DNA has also been used to include satellite DNA sequences, intergenic regions, transposable elements and "excess" DNA in high *C*-value genomes. These ideas of "selfish" DNA have not been universally accepted and the reader is referred to the extensive discussions on this fascinating theme by Cavalier-Smith (1980), Dover (1980), Smith (1980), Reid (1980), Orgel *et al*. (1980), Dover and Doolittle (1980) and Jain (1980). It is worth stressing that, even if intervening sequences have no contemporary role in gene expression, they could still be advantageous as raw material for evolving new genetic functions.

V Evolution of flanking sequences

By comparing homologous genes, it is also possible to study the evolution of 5′ and 3′ non-coding sequences found in mRNA, and of flanking non-transcribed extragenic DNA. Pairwise comparisons of the mouse and rabbit β-globin genes (Van Ooyen *et al.*, 1979), the mouse β_d^{maj}- and β_d^{min}-globin genes (Konkel *et al.*, 1979), the human β-related globin genes (Efstratiadis *et al.*, 1980), and the chicken ovalbumin, X and Y genes (Royal *et al.*, 1979; Heilig *et al.*, 1980) have shown that in general these sequences tend to evolve faster than protein coding sequences and accumulate both base substitutions and microdeletions/insertions (see Fig. 7). Detailed rates of divergence are not yet available. Nevertheless, exons tend to be recognizable as isolated regions of conserved sequences embedded within a matrix of rapidly evolving DNA. However, various conserved elements can be found in non-coding regions near genes, for example the TATA box postulated to be the promoter for RNA polymerase II (see Breathnach and Chambon, 1981) and the AATAAA sequence near the 3′ end of many eukaryotic mRNAs, which might be involved in RNA polyadenylation (Proudfoot and Brownlee, 1976). The evolution of these short sequences needs to be treated with care since they have undoubtedly evolved independently in association with many unrelated genes.

There are a few isolated reports of localized conserved sequences outside genes. For example, Takahashi *et al.* (1980) have compared clones containing the human and mouse immunoglobulin C_μ genes, and have shown a substantial conserved region of DNA on the 5′ side of the gene, perhaps involved in class switching. Similarly, Leder *et al.* (1978) have shown that close to the highly divergent mouse α- and β_d^{maj}-globin genes is a 200 bp homologous segment about 1.5 kb from the 3′ end of each gene. However, it is exceedingly unlikely that these elements are orthologous in evolution and have failed to diverge after the αβ-globin gene duplication. Instead they are perhaps members of a transposable element family which have recently moved close to both the α- and β-globin genes. It is not known whether there is any functional relation between these elements and the coordinately regulated globin genes.

Flanking regions can be recruited as new coding sequences. For example, genes for human chorionic gonadotropin (HCG), luteinizing hormone, follicle-stimulating hormone and thyroid stimulating hormone are related. However, HCG has a 30 amino acid *C*-terminal extension which appears to have evolved by loss of a termination codon and readthrough of the 3′ untranslated mRNA sequences to a UAA termination codon in the AAUAAA "polyadenylation" sequence (Fiddes and Goodman, 1980).

VI Evolution of multigene families

A Prevalence of multigene families

Early studies on amino acid sequences, chiefly of isozymes, haemoglobins and immunoglobulins, established that numerous families of proteins were coded by sets of related genes which appeared to have evolved by gene duplication and divergence. By applying recombinant DNA techniques to the study of these genes, their organization within eukaryotic genomes can be determined, and species comparisons of these families can give clues about the evolutionary history of these genes.

The first genes to be studied at the DNA level were the histone genes (see Kedes, 1979) and ribosomal RNA genes (see Long and Dawid, 1980). In both cases, multiple and essentially identical copies of these genes are arranged in tandem repetitive blocks which appear to have evolved by gene amplification by unequal crossing over between individual repeat elements, and whose current organization is thought to be an adaptation to ensure high rates of synthesis of histones or ribosomal RNAs. In this review, I shall concentrate on more recent studies of animal gene families.

The most extensive study to date has been on the globin gene family in vertebrates (see below). Numerous other examples of gene families unsuspected from protein data have recently emerged. Two genes related in sequence and structure to the single ovalbumin gene have been detected in chicken DNA by cross-hybridization with ovalbumin cDNA (Royal *et al.*, 1979; Heilig *et al.*, 1980). These X and Y genes are also expressed in the oviduct under hormonal control (LeMeur *et al.*, 1981) but their function is unknown. They are closely linked to the ovalbumin gene, and all three genes are orientated in the same direction in the order 5′-X-Y-ovalbumin-3′. These genes are separated by substantial (6 kb and 11 kb) tracts of DNA of unknown function, and have apparently evolved by a series of tandem gene duplications. Multiple copies of human interferon genes also exist. At least eight distinct (non-allelic) leukocyte interferon genes have been isolated from a human genomic DNA library, by cross-hybridization with a cloned interferon cDNA (Nagata *et al.*, 1980). At least some of these genes are clustered in human DNA, and may code for distinct interferons (Streuli *et al.*, 1980). In contrast, only one human fibroblast interferon gene has been detected (Houghton *et al.*, 1981). Comparison of leukocyte and fibroblast interferon cDNA sequences has shown that these proteins are also related but are very substantially diverged (Taniguchi *et al.*, 1980). It is likely that the interferon gene family will prove to be as ancient as, and at least as complex as, the globin family. Other cases of multigene families studied at the DNA level include at least four vitellogenin genes in *X. laevis* (Wahli *et al.*, 1979, 1980), at least two chicken δ-crystallin genes (Bhat *et al.*, 1980; Jones *et al.*, 1980), two rat preproinsulin genes (Lomedico *et al.*, 1979), three unlinked larval serum protein genes in *Drosophila* (Smith *et al.*, 1981) and multiple genes for actin in *Dictyostelium discoideum* (Kindle and Firtel, 1978; McKeown *et al.*, 1978; Firtel *et al.*, 1979; Vandekerckhove and Weber, 1980), *Drosophila* (Fyrberg *et al.*, 1980; Tobin *et al.*, 1980) and the sea urchin *Strongylocentrotus purpuratus* (Durica *et al.*, 1980). The chorion proteins are another example of a complex family (Weldon Jones and Kafatos, 1980; Spradling *et al.*, 1980).

The most complex multigene family partially characterized to date is that coding for immunoglobulins (see Adams, 1980; Leder, 1981). There are three unlinked loci in the mouse: one containing several V_λ and two C_λ genes, a second containing 100–600 V_κ and one C_κ gene and a third containing 70–400 V_H and 8 C_H genes. At least some of these genes are arranged in tandem and separated by large spacers of unknown function. It is likely that the entire immunoglobulin family evolved by a series of tandem gene duplications and were separated into three linkage groups by processes such as chromosome duplication and translocation (see Sakano, 1979a, b; Rogers, 1980; Rogers *et al.*, 1980).

In contrast to this abundance of multigene families, either clustered or dispersed throughout the genome, there are relatively few instances of true single copy genes. Possible examples of isolated genes include the chicken conalbumin gene (Cochet *et al.*, 1979) and the human and chicken insulin genes (Bell *et al.*, 1980; Perler *et al.*, 1980). Clearly, gene duplication and divergence have been major forces in evolution for generating new genetic functions.

B Evolution of the globin gene family

The arrangement of genes coding for the human globins has been intensively studied both by restriction endonuclease analysis of human DNA and by analysis of cloned DNA fragments containing these genes. At the moment, these genes, and their homologues in other vertebrates, provide the most detailed account of the evolutionary history of a multigene family.

Human globins are coded by two unlinked clusters of genes (see Efstratiadis *et al.*, 1980; Proudfoot *et al.*, 1980). The α-globin gene cluster on human chromosome 16 contains two embryonic ζ-globin genes and two almost identical fetal/adult α-globin genes arranged in the order $5'$-ζ-ζ-α-α-$3'$. All genes are oriented in the same direction and are separated by substantial tracts of intergenic DNA. The second cluster, on human chromosome 11, codes for the β-related globins and contains a single embryonic ε-globin gene, two very similar fetal globin genes ($^G\gamma$ and $^A\gamma$), a minor adult δ-globin gene and the major adult β-globin gene. These too are arranged in the same direction, in the order $5'$-ϵ-$^G\gamma$-$^A\gamma$-δ-β-$3'$ (see Fig. 9). A phylogeny of these α- and β-related genes derived from a comparison of gene sequences is shown in Fig. 5. These genes have clearly evolved by a series of tandem gene duplications and have at some stage become unlinked to give the α- and β-globin gene clusters. After duplication, the genes have diverged both in sequence and in developmental expression to give the current organization of developmentally regulated genes.

Only about 5% of the DNA in these clusters codes for globin; the remainder, consisting mainly of intergenic regions, has no known function, although there is evidence to suggest that at least some of these sequences are involved in the control of globin gene expression (Fritsch *et al.*, 1979; Van Der Ploeg *et al.*, 1980). The apparent scarcity of coding sequences in the α- and β-globin gene clusters, and indeed in clusters of immunoglobulin genes and ovalbumin-related genes, seems to reflect the DNA excess in higher eukaryotic genomes. Orgel and Crick (1980) have suggested that these extensive intergenic sequences, like intervening sequences, may have no function, but instead represent "selfish" DNA.

The arrangement of primate δ- and β-globin genes has been compared by restriction endonuclease mapping of genomic DNA in man, great apes and Old World monkeys, using human globin cDNAs to detect homologous primate genes (Jeffreys and Barrie, 1980; Martin *et al.*, 1980; Zimmer *et al.*, 1980). The organization of these two genes was indistinguishable in all species examined, with the exception of an additional 1 kb of DNA between the gorilla δ- and β-globin genes, reported by Zimmer *et al.* (1980) but not seen by Jeffreys and Barrie (1980).

This analysis has been extended by Barrie *et al.* (1981) to include the organization of the entire β-globin cluster in representative species of each of the major groups of primates (man, great apes, Old World monkeys, New World monkeys and prosimians). Sufficient conservation of β-related globin sequences has prevailed in primate evolution to enable human probes to detect and discriminate between the orthologous primate genes. The β-globin gene clusters are compared in Fig. 9 in relation to the approximate divergence times of the primate groups deduced from molecular studies (Sarich and Cronin, 1977). The entire cluster is indistinguishable in man, gorilla and baboon, indicating that the arrangement of the human cluster was

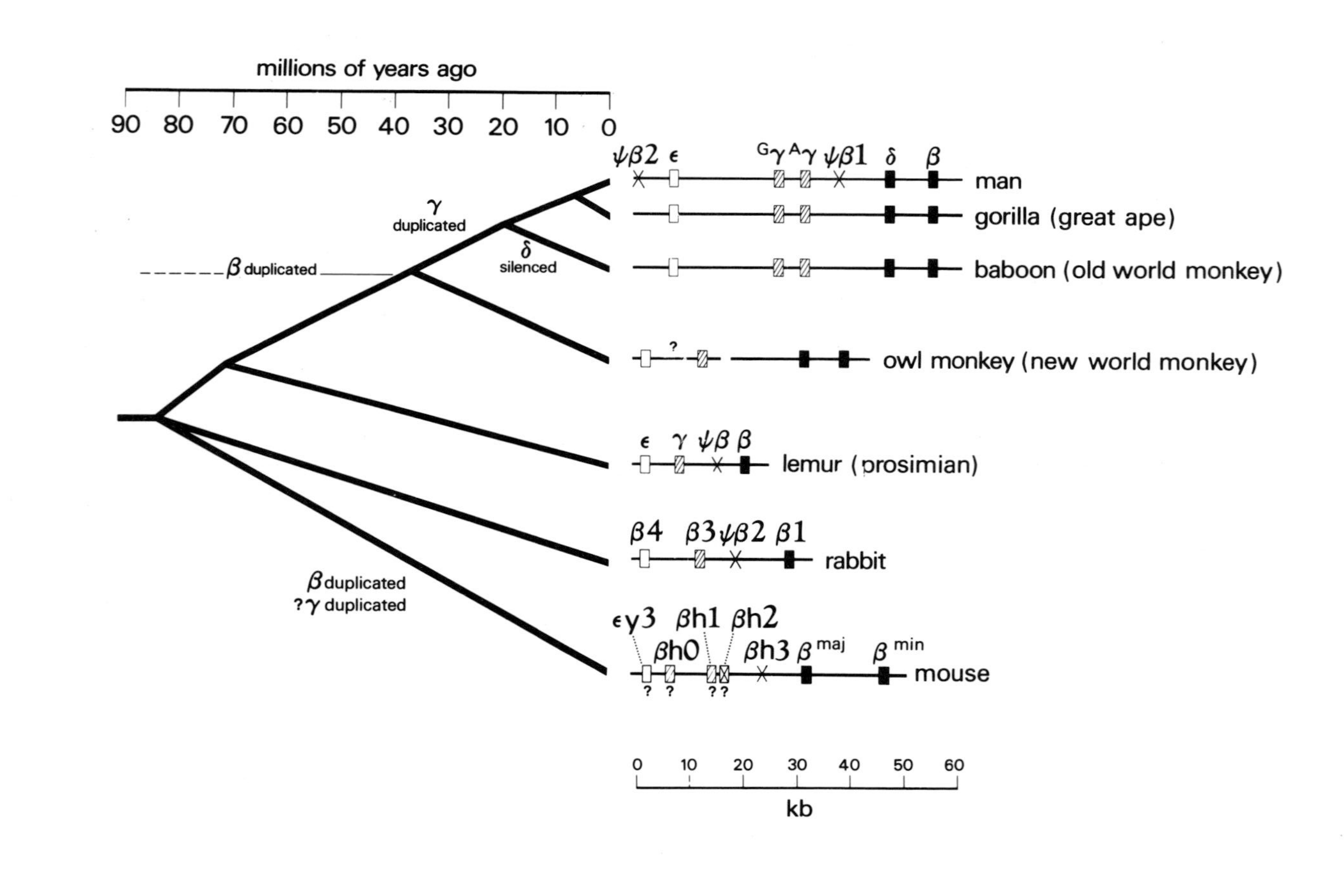

millions of years ago
90 80 70 60 50 40 30 20 10 0
γ duplicated
β duplicated
δ silenced
β duplicated
?γ duplicated
ψβ2 ε ^Gγ ^Aγ ψβ1 δ β
man
gorilla (great ape)
baboon (old world monkey)
?
owl monkey (new world monkey)
ε γ ψβ β
lemur (prosimian)
β4 β3 ψβ2 β1
rabbit
εy3 βh1 βh2
βh0 βh3 β maj β min
mouse
? ? ? ?
0 10 20 30 40 50 60
kb

established 20—40 million years ago, before the divergence of Old World monkeys and apes. The lack of rearrangement during recent primate evolution is surprising in view of the considerable number of deletions in this region detected in man and associated with haemoglobinopathies (Flavell *et al.*, 1978; Bernards *et al.*, 1979; Fritsch *et al.*, 1979; Tuan *et al.*, 1979; Bernards and Flavell, 1980; Van Der Ploeg *et al.*, 1980). In addition, the degree of restriction site divergence between man, gorilla and baboon suggests that the intergenic DNA has evolved slowly with a UEP of 5, compared with UEP = 10 for replacement site substitutions and UEP = 1.3 for silent site changes (Efstratiadis *et al.*, 1980). The apparent conservation of the arrangement and sequence of intergenic DNA in the β-globin gene cluster implies function and appears to be incompatible with the suggestion that this might be "selfish" DNA.

The duplicated $\delta\beta$-globin gene seems to have arisen about 40—70 million years ago (Fig. 9), consistent with the divergence time estimated from DNA sequence comparisons (Fig. 5). The δ-globin polypeptide is found in New World monkeys, apes and man but not in Old World monkeys (Boyer *et al.*, 1969, 1971), even though the δ-globin gene is present. Martin *et al.* (1980) have suggested that the δ-globin gene has become completely silent recently in Old World monkeys. The γ-globin gene duplication in man arose about 20—40 million years ago, and is present in Old World monkeys but not New World monkeys or prosimians. In sharp contrast, DNA sequence

---

Figure 9 Phylogeny of the β-globin gene cluster in mammals. The arrangement of the human cluster was determined by Fritsch *et al.* (1980), the primate clusters by Barrie *et al.* (1981), the rabbit cluster by Lacy *et al.* (1979) and the mouse cluster by Jahn *et al.* (1980). These maps show genes which are probably expressed (boxes) and known pseudogenes (crosses). All genes are transcribed from left to right, and intervening sequences within these genes are not shown. ϵ-related genes are shown by open boxes, γ-related genes by hatched boxes and adult β-globin genes by solid boxes (Barrie *et al.*, 1981). Orthologies with the mouse embryonic genes have not yet been determined, and the assignment of $\epsilon y3$ to ϵ and the closely-related $\beta h0$ and $\beta h1$ to γ is completely speculative. Mouse $\beta h2$ and $\beta h3$ are highly divergent, and $\beta h3$, at least, is a pseudogene. Genes orthologous to human $\psi\beta1$ and $\psi\beta2$ are probably present at equivalent positions in the gorilla and baboon. The linkage of owl monkey ϵ- and γ-globin genes has not been firmly established, and the linkage of ϵ, γ to δ, β is unknown. These maps are arranged phylogenetically using divergence times cited by Dayhoff (1972), Romero-Herrera *et al.* (1973) and Sarich and Cronin (1977).

comparison of the $^{G}\gamma$- and $^{A}\gamma$-globin genes in man gives a divergence time of much less than one million years (Efstratiadis *et al.*, 1980; Slightom *et al.*, 1980; see Fig. 5). This failure to diverge is an example of concerted evolution (see below).

The β-globin gene cluster in the lemur (a prosimian) is the shortest reported in mammals to date, and contains single ϵ-, γ- and β-globin genes (Barrie *et al.*, 1981). A similar, though longer, cluster has been cloned from the rabbit and shown to contain probably three expressed globin genes, termed $\beta 1$, $\beta 3$ and $\beta 4$ (Lacy *et al.*, 1979; see Fig. 9). Cross-hybridization with human globin DNA suggests that the $\beta 1$, $\beta 3$ and $\beta 4$ genes are probably orthologous to the human β-, γ- and ϵ-globin genes respectively (Barrie *et al.*, 1981). It therefore seems likely that a simple cluster similar to that in the lemur and rabbit was established at least 85 million years ago, before the radiation of the mammals (Romero-Herrera *et al.*, 1973). This antiquity of the ϵ-, γ- and β-globin genes is consistent with their times of divergence estimated from DNA sequences (Efstratiadis *et al.*, 1980; see Fig. 5).

The β-globin gene cluster has also been characterized in detail in the BALB/c mouse (Jahn *et al.*, 1980; see Fig. 9). Seven gene sequences have been identified, although at least one is a pseudogene (see below). The adult β-globin gene is duplicated to give a β_{d}^{maj}- and β_{d}^{min}-globin gene analogous to the β- and δ-globin genes in man. However, these gene pairs in man and mouse are almost certainly not orthologous in evolution, but instead are paralogues evolved by independent gene duplications from a single β-globin gene present in the common ancestor of man and mouse. By comparing partial sequences of the mouse embryonic β-related genes, three subfamilies of genes could be discerned (Jahn *et al.*, 1980): ϵy3; the closely related βh0 and βh1; and the highly divergent βh2 and pseudogene βh3. It is tempting to hypothesize that ϵy3 is orthologous to ϵ in primates, that βh0 and βh1 are duplicated γ-globin genes and that perhaps βh2 (as well as βh3) is a pseudogene. In this way, both the human and mouse β-globin gene clusters can be seen to have evolved by independent duplication of γ- and β-globin genes. In man, the 5′ β-globin gene diverged to become the minor (δ) adult globin gene; in mouse, the 3′ β-globin gene assumed this role.

During 400 million years of independent evolution of the β-globin gene cluster in mammals (Fig. 9), four independent active gene duplications have been fixed in a cluster with on average four active genes. This suggests that an active gene duplication is fixed per gene on average about once every 400 million years.

There is little information on the evolution of the human α-related globin gene cluster. Zimmer *et al.* (1980) have compared the duplicated adult α-globin gene arrangement in man, great apes and the gibbon, and find identical arrangements, except for a few small deletions and insertions in some species.

The α- and β-globin gene clusters also lie on different chromosomes in the chicken (Hughes *et al.*, 1979; Engel and Dodgson, 1980). However, the chicken β-globin gene cluster contains an embryonic gene on the 3′ side of the adult gene (Dodgson *et al.*, 1979). The relation between these genes and the mammalian β-related globin genes is unknown: birds and mammals diverged about 270 million years ago (Dayhoff, 1972) perhaps before the emergence of mammalian ε-, γ- and β-globin genes (Fig. 5), and it is entirely possible that the avian cluster has arisen by a completely independent series of β-globin gene duplications.

The α- and β-globin genes have also been analysed in *Xenopus laevis* (Jeffreys *et al.*, 1980; Patient *et al.*, 1980). The major adult haemoglobin is coded by a single α^1-globin gene and one β^1-globin gene. Unlike mammals and birds, these genes are closely linked in the order $5'$-α^1-β^1-$3'$. This arrangement strongly suggests that the initial αβ-globin gene duplication which occurred perhaps 500 million years ago was a tandem gene duplication, and that the tandem duplicates have since remained closely linked in Amphibia. In contrast, these genes became unlinked in the reptilian ancestors of birds and mammals, perhaps about 300 million years ago. This unlinking of the α- and β-globin genes might have resulted from a translocation between the genes, or alternatively by chromosome duplication to give two unlinked αβ clusters which could evolve towards the contemporary mammalian arrangement of α- and β-clusters by silencing of linked β- and α-genes. The importance of chromosome duplication and polyploidization in vertebrate evolution has been emphasized repeatedly (Ohno, 1970, 1973; Lalley *et al.*, 1978; Lundin, 1979).

Xenopus laevis possesses a second αβ-globin gene cluster which codes for minor adult α- and β-globin polypeptides (Jeffreys *et al.*, 1980). This cluster has the same arrangement as the α^1-β^1-globin gene cluster and appears to have arisen by tetraploidization in an ancestor of *X. laevis*. A contemporary equivalent of this ancestor, *X. tropicalis*, has as expected a single α- and β-globin gene arranged $5'$-α-β-$3'$. Thus chromosome duplication has generated globin diversity in *X. laevis* by creating two complete α-β-clusters; subsequent divergence seems to be silencing the linked α- and β-globin gene in only one cluster.

C Gene duplication by unequal crossing over

Gene duplication in for example the β-globin gene cluster has most probably arisen by unequal crossing over between homologous sequences with the cluster. The same mechanism is probably responsible for the generation of tandem repetitive sequences such as satellite DNAs, histone genes and ribosomal DNA. There are several clear instances in globin clusters of altered gene arrangements which have arisen by unequal crossing over. The fused $\delta\beta$-globin polypeptide in Hb Lepore has been shown, by restriction endonuclease mapping of nuclear DNA, to have arisen by unequal crossing over between the related δ- and β-globin genes to produce a fused $\delta\beta$-globin gene (Flavell *et al.*, 1978). The reciprocal chromosome bearing δ-, ($\beta\delta$-) and β-globin genes and specifying Hb P Congo (Lehmann and Charlesworth, 1970) has not been characterized at the DNA level. Chromosomes carrying three α-globin genes or a single α-globin gene, in addition to those carrying the duplicated α-locus, have been detected in man (Orkin, 1978; Embury *et al.*, 1979b; Goossens *et al.*, 1980; Higgs *et al.*, 1980) and in the chimpanzee (Zimmer *et al.*, 1980). In many cases these three-gene and one-gene chromosomes appear to be the reciprocal products of unequal crossing over between the two α-globin gene loci. The point of cross-over need not necessarily be within an α-globin gene, since each α-globin gene is located towards the 3′ end of a 4 kb region of homology (Lauer *et al.*, 1980; Proudfoot and Maniatis, 1980). For example, one α-globin gene deletion has been reported which can be interpreted as the product of unequal recombination between the 5′ region of each homology block (Embury *et al.*, 1979a). Unequal crossing over between two duplicated genes appears to have generated a cluster of three glutamate tRNA genes in *Drosophila* (Hosbach *et al.*, 1980). Unequal crossing over in the ribosomal DNA of the yeast *Saccharomyces cerevisiae* has been directly demonstrated genetically by monitoring the deletion or duplication of a *LEU*2 marker inserted within the ribosomal DNA locus (Szostak and Wu, 1980).

Numerous repetitive DNA sequence elements have been found in the human β-globin gene cluster (Coggins *et al.*, 1980; Fritsch *et al.*, 1980). Unequal recombination between similar elements flanking a single copy gene might have been responsible for the $^{G}\gamma$-, $^{A}\gamma$-globin gene duplication in Old World monkeys. If so, then the spacing between the two γ-globin genes will have been determined by the positions of repeated elements near these genes. An analogous example, where gene duplication might be caused by recombination between repetitive elements, could occur during amplification of the dihydrofolate reductase gene in mammalian cells selected for resistance to methotrexate (Nunberg *et al.*, 1980; Schimke, 1980).

VII Pseudogenes

Recent analyses of multigene families are beginning to show additional gene sequences which have accumulated mutations and no longer represent functional genes. The first example of a pseudogene was found in the tandem repetitive oocyte-type 5S DNA of *Xenopus laevis* (Jacq *et al.*, 1977). Each 700 bp repeat was found to contain a complete 121 bp 5S gene plus a 101 bp pseudogene, lacking the 3′ end of the 5S gene and showing some sequence divergence from the functional gene.

More recently, several globin pseudogenes have been reported. Low stringency hybridizations have been used to hunt for additional β-related globin sequences in the human β-globin gene cluster. A $\psi\beta 1$ globin gene was detected between the $^{A}\gamma$- and δ-globin genes, and a $\psi\beta 2$ gene to the 5′ side of the ϵ-globin gene (Fritsch *et al.*, 1980; Efstratiadis *et al.*, 1980; see Fig. 9). These sequences cannot be identified with any known β-globin polypeptides, and are probably pseudogenes. However, detailed sequences of these genes have not yet been published. The $\psi\beta 1$ and $\psi\beta 2$ sequences are probably also present in apes and Old World monkeys (Barrie *et al.*, 1981). The rabbit $\beta 3$- (γ-like) and $\beta 1$- (β-like) globin genes are also separated by a pseudoglobin gene $\psi\beta 2$ (Lacy and Maniatis, 1980; see Fig. 9). Complete sequencing has shown substantial divergence from the $\beta 1$-globin gene, including frameshift mutations, premature termination codons and disruption of normal intron/exon junction sequences sufficient to render the gene incapable of coding globin. It is not known whether the gene is transcribed. By comparing rabbit $\psi\beta 2$ and $\beta 1$ DNA sequences it was found that replacement sites had not diverged as much as silent sites, which suggests that the former sites at least were under selective pressure for some time after the $\psi\beta 2$-$\beta 1$ duplication. The simplest interpretation is that the duplication initially gave two functional genes, and that the 5′ gene was later silenced, releasing it from conservation of replacement sites. Using the (approximate) clock rates of silent site and replacement site substitution (Efstratiadis *et al.*, 1980), Lacy and Maniatis (1980) estimate that the duplication arose at least 50 million years ago, and that $\psi\beta 2$ was eventually silenced 30 million years ago. The mouse embryonic and adult β-related globin genes are also separated by at least one highly diverged pseudogene, βh3 (Jahn *et al.*, 1980; see Fig. 9). The 3′ end of this gene is closely homologous to the adult β-globin gene, except for a frameshift mutation, whereas the 5′ end of the gene is highly diverged. A similar pseudogene is seen in the lemur between the γ- and β-globin genes (Barrie *et al.*, 1981; Fig. 9). In this case, the $\psi\beta$

gene contains the 3′ end of a β-globin gene preceded by sequences apparently related to the 5′ end of the ϵ-globin gene. The goat pre-adult β^{c}-globin gene is also preceded by a β^{x} pseudogene (Cleary *et al.*, 1980).

An inactive α-globin pseudogene ($\psi\alpha 1$) has been found between the embryonic (ζ-) and adult α-globin genes in man (Proudfoot and Maniatis, 1980), and contains alterations in putative splicing signals, an initiator codon mutation and frameshift deletions. The remarkable mouse α-globin pseudogene which has lost both intervening sequences (Nishioka *et al.*, 1980; Vanin *et al.*, 1980) has already been described. The linkage arrangement between this sequence and functional α-globin genes has not been reported. A V_{κ} pseudogene unable to code for a functional immunoglobin has been described by Bentley and Rabbitts (1980).

Vanin *et al.* (1980) have suggested that pseudogenes might be involved in the control of gene expression, perhaps by diverting transcription into non-productive pathways or by encoding control RNA species. However, it seems more likely that these genes are the evolutionary relics of duplicated genes which have become silenced and have been released from evolutionary constraint. Their prevalence in α- and β-globin loci suggests that excessive duplication might not be a rare event within gene clusters. Of course, the pseudogenes detected so far all show a reasonable (> 70%) homology with functional globin genes. More diverged genes would not be detectable by hybridization with functional gene sequences and could only be located by laborious computer searches of extragenic DNA sequences. It seems highly likely that a continuous spectrum of pseudogenes exists, ranging from very recently silenced genes (such as the Old World monkey δ-globin gene?) to sequences so diverged that there is no residual homology whatsoever. The proportion of intergenic DNA occupied by highly diverged pseudogenes is not known.

VIII Concerted evolution

When a gene duplicates, each duplicate locus does not necessarily diverge independently during evolution. Several instances have emerged where duplicated genes instead appear to interchange sequences by some mechanism which maintains a close sequence homology between the duplicated loci. Zimmer *et al.* (1980) have called this process "concerted evolution".

The human ${}^{G}\gamma$- and ${}^{A}\gamma$-globin genes give the clearest example of concerted evolution. The γ-globin gene duplication arose 20—40 million years ago and is currently present in at least some Old World monkeys and great apes, as well as in man (Barrie *et al.*, 1981). Jeffreys (1979) described a restriction endonuclease cleavage site polymorphism in the second intervening sequence of the human $^{G}\gamma$-globin gene which was also present at the same position in the $^{A}\gamma$-globin gene, and postulated that a mechanism such as intra-chromosomal recombination between these loci existed to move this variant from one γ-globin gene to another. The $^{G}\gamma$- and $^{A}\gamma$-globin genes from a single human chromosome have been sequenced by Slightom *et al.* (1980). The 5′ two-thirds of these genes were almost identical ($>$ 99% homologous) whereas the 3′ ends of these genes, including part of the second intervening sequence, showed more substantial differences particularly in non-coding regions. At the homology junction in the intervening sequence was a tract of simple sequence DNA consisting mainly of $(TG)_n$. It is suggested that this sequence is a hot spot for recombination between mis-aligned $^{G}\gamma$- and $^{A}\gamma$-globin genes either in a single DNA molecule, or between sister chromatids, or between identical chromosomes in a homozygote. During recombination, branch migration towards the 5′ side of this sequence can lead to gene conversion of the 5′ region of one or other γ-globin gene (Fig. 10), to give the localized region of homology in the $^{G}\gamma$- and $^{A}\gamma$-globin genes. Slightom *et al.* (1980) also compared two allelic $^{A}\gamma$-globin genes. Surprisingly, the 5′ region of the two $^{A}\gamma$ alleles differed more than did the linked $^{G}\gamma$- and $^{A}\gamma$-globin genes, whereas the converse was true for the 3′ region. These localized differences can also be accommodated by more orthodox interallelic conversions.

The two α-globin genes in man also appear to be evolving in concert. Zimmer *et al.* (1980) have compared these genes in man and great apes, and show that the α-globin gene duplication arose at least 8 million years ago. They also suggest from protein studies that the duplicated state might have been in existence for as long as 300 million years; however, there is no evidence that for example the chicken and human α-globin gene duplicates are orthologous, rather than being products of independent gene duplications such as the $\delta\beta$-globin genes of man and $\beta_d^{maj}\beta_d^{min}$-globin genes of mouse. Nevertheless, the discrepancy between the age of the α duplication and the extremely similarity of the α-globin genes in man suggests genetic interchange between these loci. One mechanism, other than gene conversion, is suggested by the common occurrence of three-α and one-α loci in man, resulting from unequal recombination between these loci. Multiple rounds of unequal crossing over between diverged α-globin genes accompanied by fixation of chromosomes now containing identical genes can result in concerted evolution (see Zimmer *et al.*, 1980). This model is analogous to one of the

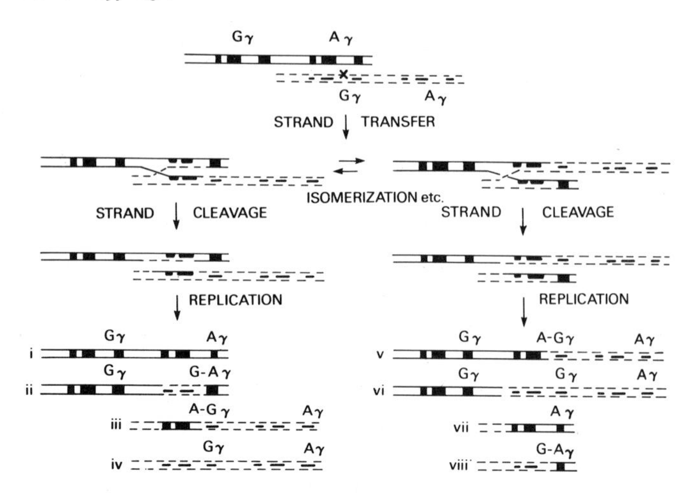

Figure 10 A model for gene conversion between the human $^{G}\gamma$- and $^{A}\gamma$-globin genes. Coding sequences are shown as black bars in two misaligned sister chromatids represented by continuous and dashed lines. Recombination is initiated at the simple sequence hotspot (X) in the second intervening sequence. Branch migration and isomerization occur, followed by strand cleavage and ligation. Replication of the non-isomerized complex yields two types of duplicated γ loci showing interlocus gene conversion (ii, iii), whereas replication of the isomer generates three-γ and one-γ chromosomes (v—viii). From Slightom *et al.* (1980), reprinted with permission of the MIT press.

mechanisms proposed for the maintenance of sequence homogeneity within families of repeated sequences (see Tartof, 1975). This mechanism seems less likely for the γ-globin genes since no three-γ or one-γ loci have yet been detected in man, despite extensive screening (Little *et al.*, 1979).

Concerted evolution by gene conversion or mutiple unequal crossing over can also be invoked to account for the lack of divergence of the first intervening sequence in the mouse β_d^{maj}- and β_d^{min}-globin genes (Konkel *et al.*, 1979), the relative similarity of the 3′ ends of the mouse βh3 pseudogene and the β_d^{maj}-globin gene (Jahn *et al.*, 1980), the excessive divergence of the 3′ ends of the human β- and δ-globin genes (Efstratiadis *et al.*, 1980), and what appears to be an ϵ—β hybrid structure of the lemur pseudogene (Barrie *et al.*, 1981). In the latter case, genetic exchange might therefore have occurred between non-adjacent genes within a cluster.

Concerted evolution has also been described in the heat shock genes of *Drosophila*. The major heat shock protein (hsp70) is coded by 4—8 genes arranged in two linked clusters. In both clusters, inverted repeats containing hsp70 genes can be found. This head-to-head arrangement is unusual, and contrasts to the tandem arrays seen in most gene clusters. Oppositely orientated genes have also been noted in the *Drosophila* histone gene cluster, but not in that of sea urchins (see Kedes, 1979). Leigh Brown and Ish-Horowicz (1981) have compared both clusters of hsp70 genes by the restriction endonuclease mapping of DNA from three *Drosophila* species. The arrangement of both clusters was the same in all three species, except for a substantial insertion in one cluster in *Drosophila melanogaster*. By comparing restriction sites within the hsp70 genes, it was seen that an evolutionary change in a cleavage site in one gene was often accompanied by a corresponding change in other hsp70 genes. Thus a fixed variant at one locus has spread to other loci, resulting in the concerted evolution of hsp70 genes, both between genes in an inverted repeat and between the two clusters. They also suggest that sequence homogeneity is maintained by interlocus gene conversion. Similarly, Arnheim *et al.* (1980) have compared human and great ape ribosomal DNA, and have shown that these repeated genes are more homogeneous within than between species, despite the fact that this multigene family is scattered over several non-homologous chromosomes. Specific metaphase associations between these chromosomes have been observed, and may be responsible for mediating genetic interchange (Miller *et al.*, 1977).

Gene conversion in yeast has been directly demonstrated between unlinked genes (Hofer *et al.*, 1979; Scherer and Davis, 1980). Frequent meiotic conversions between linked genes have also been shown in a *S. cerevisiae* strain carrying a tandem duplication of the *LEU*2 gene (Klein and Petes, 1981). This provides direct evidence for the mechanism postulated by Slightom *et al.* (1980) for the intrachromosomal conversion of the $^{G}\gamma$- and $^{A}\gamma$-globin genes (Fig. 10).

IX Future prospects

It is now a little over three years since a single copy gene was first cloned from a higher organism. Since then, most research has been directed towards an understanding of gene structure and function. Yet these analyses have repeatedly moved into the field of molecular evolution. New conjectures from this research are leading us back to the earliest stages of the evolution of life, when primitive organisms perhaps used RNA splicing to shuffle and reshuffle their limited repertoire of inherited information in an attempt to adapt more efficiently to ever-changing environments. More recent evolutionary histories of genes and gene clusters are emerging, including the first steps towards a detailed understanding of the globin dynasty which has spanned a period of more than 1000 million years. The first estimates of rates of DNA sequence evolution are appearing, plus a variety of evolutionary mechanisms for duplicating genes, unlinking genes, silencing them or preventing duplicates from evolving independently. For three years of research, most of it not directly concerned with evolutionary questions, the progress is impressive. What of the future?

The first major battle is going to develop between neutralists and selectionists over the possible existence of evolutionary DNA clocks and the mechanisms that run them. The preliminary results (Perler *et al.*, 1980; Efstratiadis *et al.*, 1980) are tantalizing but insufficient for any firm debate. Is there a silent site substitution clock, and is it driven by neutral drift? Could such a fast running clock be used to construct reliable phylogenies of closely related species? What of the extensive sequences between genes? Are they just an accumulation of junk and pseudogenes or are they really functional? How do these sequences evolve? Only time and a great deal of DNA sequencing will give the answers.

All of these studies concentrate on minute areas of the genome. Yet we know that gross chromosome rearrangements are also important in speciation and evolution. Doubtless, a direct molecular analysis of chromosomal translocations, deletions, insertions and inversions will reveal the mechanisms behind these events, and may well implicate elements such as transposable sequences as key agents in karyotypic evolution. In addition, transposable DNA sequences deserve a great deal more study, to determine their ubiquity and indeed their very existence in eukaryotes, their possible role in gene duplication and dispersal, and the evolutionary relationships between these sequences and extrachromosomal genetic elements such as retroviruses (see Temin, 1980).

What molecular events underlie major morphological changes in evolution? Why are men and great apes so different whilst their DNA remains almost identical in sequence (King and Wilson, 1975)? Are there a limited number of key regulatory differences, or instead do a multitude of small changes each with an infinitesimal effect on morphology, together generate the phenotype differences? Could one identify such sequences by using recombinant DNA technology to create hybrid species containing specific replacements of DNA regions of interest? Time alone will tell.

X Acknowledgements

I am grateful to Dr R. Semeonoff and Paul Barrie for many helpful discussions and critical comments on the manuscript.

XI References

Abelson, J. (1979). *A. Rev. Biochem.* **48**, 1035–1069.

Adams, J. M. (1980). *Immunol. Today* **1**, 10–17.

Appleby, C. A. (1974). *In* "The Biology of Nitrogen Fixation" (Ed. A. Quispel) 521–554. North Holland, Amsterdam.

Arnheim, N., Krystal, M., Schmickel, R., Wilson, G., Ryder, O. and Zimmer, E. (1980). *Proc. Natn. Acad. Sci. U.S.A.* **77**, 7323–7327.

Baralle, F. E., Shoulders, C. C. and Proudfoot, N. J. (1980). *Cell* **21**, 621–626.

Barrie, P. A., Jeffreys, A. J. and Scott, A. F. (1981). *J. Mol. Biol.*, in press.

Bell, G. I., Pictet, R. L., Rutter, W. J., Cordell, B., Tischer, E. and Goodman, H. M. (1980). *Nature, Lond.* **284**, 26–32.

Bentley, D. L. and Rabbitts, T. H. (1980). *Nature, Lond.* **288**, 730–733.

Bernards, R. and Flavell, R. A. (1980). *Nucl. Acids Res.* **8**, 1521–1534.

Bernards, R., Kooter, J. M. and Flavell, R. A. (1979). *Gene* **6**, 265–280.

Bhat, S. P., Jones, R. E., Sullivan, M. A. and Piatigorsky, J. (1980). *Nature, Lond.* **284**, 234–238.

Bird, A. and Southern, E. (1978). *J. Mol. Biol.* **118**, 27–48.

Blake, C. C. F. (1979). *Nature, Lond.* **277**, 598.

Bodmer, W. F. (1981). *Am. J. Human Genet.*, in press.

Bonner, T. I., Heinemann, R. and Todaro, G. J. (1980). *Nature, Lond.* **286**, 420–423.

Borst, P. and Grivell, L. A. (1981). *Nature, Lond.* **289**, 439–440.

Boyer, S. H., Crosby, E. F., Thurmon, T. F., Noyes, A. N., Fuller, G. F., Leslie, S. E., Shepard, M. K. and Herndon, C. N. (1969). *Science* **166**, 1428–1431.

Boyer, S. H., Crosby, E. F., Noyes, A. N., Fuller, G. F., Leslie, S. E., Donaldson, L. J., Vrablik, G. R., Schaefer, E. W. and Thurmon, T. F. (1971). *Biochem. Genet.* **5**, 405–448.

Breathnach, R. and Chambon, P. (1981). *A. Rev. Biochem.* **50**, in press.

Cavalier-Smith, T. (1980). *Nature, Lond.* **285**, 617–618.

Chang, A. C. Y., Cochet, M. and Cohen, S. N. (1980). *Proc. Natn. Acad. Sci. U.S.A.* **77**, 4890–4894.

Chien, Y.-H. and Thompson, E. B. (1980). *Proc. Natn. Acad. Sci. U.S.A.* **77**, 4583–4587.

Cleary, M. L., Haynes, J. R., Schon, E. A. and Lingrel, J. B. (1980). *Nucl. Acids Res.* **8**, 4791–4802.

Cochet, M., Gannon, F., Hen, R., Maroteaux, L., Perrin, F. and Chambon, P. (1979). *Nature, Lond.* **282**, 567–574.

Coggins, L. W., Grindlay, G. J., Vass, J. K., Slater, A. A., Montague, P., Stinson, M. A. and Paul, J. (1980). *Nucl. Acids Res.* **8**, 3319–3333.

Craik, C. S., Buchman, S. R. and Beychok, S. (1980). *Proc. Natn. Acad. Sci. U.S.A.* **77**, 1384–1388.

Crick, F. (1979). *Science* **204**, 264–271.

Darnell, J. E. (1978). *Science* **202**, 1257–1260.

Dayhoff, M. O. (Ed.)(1972). "Atlas of Protein Sequence and Structure", Vol. 5. Natn. Biomed. Res. Found., Silver Spring, Md.

Dodgson, J. B., Strommer, J. and Engel, J. D. (1979). *Cell* **17**, 879–887.

Doolittle, W. F. (1978). *Nature, Lond.* **272**, 581–582.

Doolittle, W. F. and Sapienza, C. (1980). *Nature, Lond.* **284**, 601–603.

Dover, G. (1980). *Nature, Lond.* **285**, 618–620.

Dover, G. and Doolittle, W. F. (1980). *Nature, Lond.* **288**, 646–647.

Durica, D. S., Schloss, J. A. and Crain, W. R. (1980). *Proc. Natn. Acad. Sci. U.S.A.* **77**, 5683–5687.
Early, P. W., Davis, M. M., Kaback, D. B., Davidson, N. and Hood, L. (1979). *Proc. Natn. Acad. Sci. U.S.A.* **76**, 857–861.
Early, P., Rogers, J., Davis, M., Calame, K., Bond, M., Wall, R. and Hood, L. (1980). *Cell* **20**, 313–319.
Eaton, W. A. (1980). *Nature, Lond.* **284**, 183–185.
Efstratiadis, A., Kafatos, F. C. and Maniatis, T. (1977). *Cell* **10**, 571–585.
Efstratiadis, A., Posakony, J. W., Maniatis, T., Lawn, R. M., O'Connell, C., Spritz, R. A., DeRiel, J. K., Forget, B. G., Weissman, S. M., Slightom, J. L., Blechl, A. E., Smithies, O., Baralle, F. E., Shoulders, C. C. and Proudfoot, N. J. (1980). *Cell* **21**, 653–668.
Embury, S. H., Lebo, R. V., Dozy, A. M. and Kan, Y. W. (1979a). *J. Clin. Invest.* **63**, 1307–1310.
Embury, S. H., Miller, J., Chan, V., Todd, D., Dozy, A. M. and Kan, Y. W. (1979b). *Blood* **54** (Suppl.), 53a.
Engel, J. D. and Dodgson, J. B. (1980). *Proc. Natn. Acad. Sci. U.S.A.* **77**, 2596–2600.
Fiddes, J. C. and Goodman, H. M. (1980). *Nature, Lond.* **286**, 684–687.
Firtel, R. A., Timm, R., Kimmel, A. R. and McKeown, M. (1979). *Proc. Natn. Acad. Sci. U.S.A.* **76**, 6206–6210.
Flavell, R. A., Kooter, J. M., De Boer, E., Little, P. F. R. and Williamson, R. (1978). *Cell* **15**, 25–41.
Francki, R. I. B. and Randles, J. W. (1980). *In* "Rhabdoviruses" (Ed. D. H. L. Bishop) Vol. 3. CRC Press, Boca Raton, Fla.
Fritsch, E. F., Lawn, R. M. and Maniatis, T. (1979). *Nature, Lond.* **279**, 598–603.
Fritsch, E. F., Lawn, R. M. and Maniatis, T. (1980). *Cell* **19**, 959–972.
Fyrberg, E. A., Kindle, K. L., Davidson, N. and Sodja, A. (1980). *Cell* **19**, 365–378.
Gale, J. S. (1980). "Population Genetics." Blackie, Glasgow, London.
Gallwitz, D. and Sures, I. (1980). *Proc. Natn. Acad. Sci. U.S.A.* **77**, 2546–2550.
Gannon, F., O'Hare, K., Perrin, F., LePennec, J. P., Benoist, C., Cochet, M., Breathnach, R., Royal, A., Garapin, A., Cami, B. and Chambon, P. (1979). *Nature, Lond.* **278**, 428–434.
Gilbert, W. (1978). *Nature, Lond.* **271**, 501.
Goossens, M., Dozy, A. M., Embury, S. H., Zachariades, Z., Hadjiminas, M. G., Stamatoyannopoulos, G. and Kan, Y. W. (1980). *Proc. Natn. Acad. Sci. U.S.A.* **77**, 518–521.
Grantham, R., Gautier, C. and Gouy, M. (1980a). *Nucl. Acids Res.* 8, 1893–1912.
Grantham, R., Gautier, C., Gouy, M., Mercier, R. and Pavé, A. (1980b). *Nucl. Acids Res.* 8, r49–r62.
Grantham, R., Gautier, C., Gouy, M., Jacobzone, M. and Mercier, R. (1981). *Nucl. Acids Res.* **9**, r43–r74.
Grunstein, M., Schedl, P. and Kedes, L. (1976). *J. Mol. Biol.* **104**, 351–369.
Hagenbüchle, O., Tosi, M., Schibler, U., Bovey, R., Wellauer, P. K. and Young, R. A. (1981). *Nature, Lond.* **289**, 643–646.
Hamer, D. H. and Leder, P. (1979a). *Cell* **17**, 737–747.
Hamer, D. H. and Leder, P. (1979b). *Cell* 18, 1299–1302.
Hardison, R. C., Butler, E. T., Lacy, E., Maniatis, T., Rosenthal, N. and Efstratiadis, A. (1979). *Cell* 18, 1285–1297.

Harris, H. (1980). "The Principles of Human Biochemical Genetics", 3rd edn. Elsevier, Amsterdam.
Harris, H. and Hopkinson, D. A. (1972). *Ann. Hum. Genet.* **36**, 9–19.
Heilig, R., Perrin, F., Gannon, F., Mandel, J. L. and Chambon, P. (1980). *Cell* **20**, 625–637.
Higgs, D. R., Old, J. M., Pressley, L., Clegg, J. B. and Weatherall, D. J. (1980). *Nature, Lond.* **284**, 632–635.
Hofer, F., Hollensten, H., Janner, F., Minet, M., Thuriaux, P. and Leupold, U. (1979). *Curr. Genet.* **1**, 45–61.
Hood, L., Campbell, J. H. and Elgin, S. C. R. (1975). *A. Rev. Genet.* **9**, 305–353.
Hosbach, H. A., Silberklang, M. and McCarthy, B. J. (1980). *Cell* **21**, 169–178.
Houghton, M., Jackson, I. J., Porter, A. G., Doel, S. M., Catlin, G. H., Barber, C. and Carey, N. H. (1981). *Nucl. Acids Res.* **9**, 247–266.
Hughes, S. H., Stubblefield, E., Payvar, F., Engel, J. D., Dodgson, J. B., Spector, D., Cordell, B., Schimke, R. T. and Varmus, H. E. (1979). *Proc. Natn. Acad. Sci. U.S.A.* **76**, 1348–1352.
Jacq, C., Miller, J. R. and Brownlee, G. G. (1977). *Cell* **12**, 109–120.
Jahn, C. L., Hutchinson, C. A., Phillips, S. J., Weaver, S., Haigwood, N. L., Voliva, C. F. and Edgell, M. H. (1980). *Cell* **21**, 159–168.
Jain, H. K. (1980). *Nature, Lond.* **288**, 647–648.
Jeffreys, A. J. (1979). *Cell* **18**, 1–10.
Jeffreys, A. J. and Barrie, P. A. (1980). *Phil. Trans. R. Soc. Ser. B*, in press.
Jeffreys, A. J., Wilson, V., Wood, D., Simons, J. P., Kay, R. M. and Williams, J. G. (1980). *Cell* **21**, 555–564.
Jones, R. E., Bhat, S. P., Sullivan, M. A. and Piatigorsky, J. (1980). *Proc. Natn. Acad. Sci. U.S.A.* **77**, 5879–5883.
Jukes, T. H. (1980). *Science* **210**, 973–978.
Jukes, T. H. and King, J. L. (1979). *Nature, Lond.* **281**, 605–606.
Kafatos, F. C., Efstratiadis, A., Forget, B. G. and Weissmann, S. M. (1977). *Proc. Natn. Acad. Sci. U.S.A.* **74**, 5618–5622.
Kedes, L. H. (1979). *A. Rev. Biochem.* **48**, 837–870.
Kimura, M. (1969). *Proc. Natn. Acad. Sci. U.S.A.* **63**, 1181–1188.
Kimura, M. (1977). *Nature, Lond.* **267**, 275–276.
Kindle, K. and Firtel, R. A. (1978). *Cell* **15**, 763–778.
King, J. L. and Jukes, T. H. (1969). *Science* **164**, 788–798.
King, M.-C. and Wilson, A. C. (1975). *Science* **188**, 107–116.
Klein, H. L. and Petes, T. D. (1981). *Nature, Lond.* **289**, 144–148.
Kohne, D. E. (1970). *Q. Rev. Biophys.* **3**, 327–375.
Konkel, D. A., Maizel, J. V. and Leder, P. (1979). *Cell* **18**, 865–873.
Lacy, E. and Maniatis, T. (1980). *Cell* **21**, 545–553.
Lacy, E., Hardison, R. C., Quon, D. and Maniatis, T. (1979). *Cell* **18**, 1273–1283.
Lai, E. C., Woo, S. L. C., Dugaiczyk, A. and O'Malley, B. W. (1979). *Cell* **16**, 201–211.
Lalley, P. A., Minna, J. D. and Francke, U. (1978). *Nature, Lond.* **274**, 160–162.
Lauer, J., Shen, C. K. J. and Maniatis, T. (1980). *Cell* **20**, 119–130.
Lawn, R. M., Efstratiadis, A., O'Connell, C. and Maniatis, T. (1980). *Cell* **21**, 647–651.
Lazowska, J., Jacq, C. and Slonimski, P. P. (1980). *Cell* **22**, 333–348.
Leder, P. (1981). *A. Rev. Biochem.* **50**, in press.
Leder, A., Miller, H. I., Hamer, D. H., Seidman, J. G., Norman, B., Sullivan, M. and Leder, P. (1978). *Proc. Natn. Acad. Sci. U.S.A.* **75**, 6187–6191.
Leder, P., Hansen, J. N., Konkel, D., Leder, A., Nishioka, Y. and Talkington, C. (1980). *Science* **209**, 1336–1342.

Lehmann, H. and Charlesworth, D. (1970). *Biochem. J.* **119**, 43P.
Leigh Brown, A. J. and Ish-Horowicz, D. (1981). *Nature, Lond.*, in press.
LeMeur, M., Glanville, N., Mandel, J. L., Gerlinger, P., Palmiter, R. and Chambon, P. (1981). *Cell*, in press.
Lewin, B. (1974). "Gene Expression", Vol. 2. Wiley, London.
Lewin, B. (1980). *Cell* **22**, 324–326.
Liebhaber, S. A., Goossens, M. J. and Kan, Y. W. (1980). *Proc. Natn. Acad. Sci. U.S.A.* in press.
Little, P. F. R., Williamson, R., Annison, G., Flavell, R. A., De Boer, E., Bernini, L. F., Ottolenghi, S., Saglio, G. and Mazza, U. (1979). *Nature, Lond.* **282**, 316–318.
Lomedico, P., Rosenthal, N., Efstratiadis, A., Gilbert, W., Kolodner, R. and Tizard, R. (1979). *Cell* **18**, 545–558.
Long, E. O. and Dawid, I. B. (1980). *A. Rev. Biochem.* **49**, 727–764.
Lundin, L.-G. (1979). *Clin. Genet.* **16**, 72–81.
McKeown, M., Taylor, W. C., Kindle, K. L., Firtel, R. A., Bender, W. and Davidson, N. (1978). *Cell* **15**, 789–800.
Marcker, K. (1980), cited in *Nature, Lond.* **288**, 215–218.
Martin, S. L., Zimmer, E. A., Kan, Y. W. and Wilson, A. C. (1980). *Proc. Natn. Acad. Sci. U.S.A.* **77**, 3563–3566.
Maxam, A. M. and Gilbert, W. (1980). *In* "Methods in Enzymology" (Eds L. Grossman and K. Modare) Vol. 65, 499–560. Academic Press, New York.
Maxwell, I. H., Maxwell, F. and Hahn, W. E. (1980). *Nucl. Acids Res.* **8**, 5875–5894.
Miller, D. A., Tantravahi, R., Dev, V. G. and Miller, O. J. (1977). *Am. J. Hum. Genet.* **29**, 490–502.
Nagata, S., Mantei, N. and Weissmann, C. (1980). *Nature, Lond.* **287**, 401–408.
Neel, J. V., Mohrenweiser, H. W. and Meisler, M. H. (1980). *Proc. Natn. Acad. Sci. U.S.A.* **77**, 6037–6041.
Nei, M. and Li, W.-H. (1979). *Proc. Natn. Acad. Sci. U.S.A.* **76**, 5269–5273.
Ng, R. and Abelson, J. (1980). *Proc. Natn. Acad. Sci. U.S.A.* **77**, 3912–3916.
Nishioka, Y. and Leder, P. (1979). *Cell* **18**, 875–882.
Nishioka, Y., Leder, A. and Leder, P. (1980). *Proc. Natn. Acad. Sci. U.S.A.* **77**, 2806–2809.
Nobrega, F. G. and Tzagoloff, A. (1980). *J. Biol. Chem.* **255**, 9821–9837.
Nunberg, J. H., Kaufman, R. J., Chang, A. C. Y., Cohen, S. N. and Schimke, R. T. (1980). *Cell* **19**, 355–364.
Ohno, S. (1970). "Evolution by Gene Duplication." Springer-Verlag, Heidelberg.
Ohno, S. (1973). *Nature, Lond.* **244**, 259–262.
Ordahl, C. P., Tilghman, S. M., Ovitt, C., Fornwald, J. and Largen, M. T. (1980). *Nucl. Acids Res.* **8**, 4989–5005.
Orgel, L. E. and Crick, F. H. C. (1980). *Nature, Lond.* **284**, 604–606.
Orgel, L. E., Crick, F. H. C. and Sapienza, C. (1980). *Nature, Lond.* **288**, 645–646.
Orkin, S. H. (1978). *Proc. Natn. Acad. Sci. U.S.A.* **75**, 5950–5954.
Patient, R. K., Elkington, J. A., Kay, R. M. and Williams, J. G. (1980). *Cell* **21**, 565–573.
Perler, F., Efstratiadis, A., Lomedico, P., Gilbert, W., Kolodner, R. and Dodgson, J. (1980). *Cell* **20**, 555–566.
Proudfoot, N. J. and Baralle, F. (1979). *Proc. Natn. Acad. Sci. U.S.A.* **76**, 5435–5439.
Proudfoot, N. J. and Brownlee, G. G. (1976). *Nature, Lond.* **263**, 211–214.
Proudfoot, N. J. and Maniatis, T. (1980). *Cell* **21**, 537–545.

Proudfoot, N. J., Shander, M. H. M., Manley, J. L., Gefter, M. L. and Maniatis, T. (1980). *Science* **209**, 1329–1336.
Reanney, D. (1979). *Nature, Lond.* **277**, 598–600.
Reid, R. A. (1980). *Nature, Lond.* **285**, 620.
Rogers, J. (1980). *New Scientist* **86**, 155–157.
Rogers, J., Early, P., Carter, C., Calame, K., Bond, M., Hood, L. and Wall, R. (1980). *Cell* **20**, 303–312.
Romero-Herrera, A. E., Lehmann, H., Joysey, K. A. and Friday, A. E. (1973). *Nature, Lond.* **246**, 389–395.
Royal, A., Garapin, A., Cami, B., Perrin, F., Mandel, J. L., LeMeur, M., Brégégègre, F., Gannon, F., LePennec, J. P., Chambon, P. and Kourilsky, P. (1979). *Nature, Lond.* **279**, 125–132.
Sakano, H., Hüppi, K., Heinrich, G. and Tonegawa, S. (1979a). *Nature, Lond.* **280**, 288–294.
Sakano, H., Rogers, J. H., Hüppi, K., Brack, C., Trannecker, A., Maki, R., Wall, R. and Tonegawa, S. (1979b). *Nature, Lond.* **277**, 627–633.
Sanger, F., Nicklin, S. and Coulson, A. R. (1978). *Proc. Natn. Acad. Sci. U.S.A.* **74**, 5463–5467.
Sarich, V. M. and Cronin, J. E. (1977). *Nature, Lond.* **269**, 354.
Schafer, M. P., Boyd, C. D., Tolstoshev, P. and Crystal, R. G. (1980). *Nucl. Acids Res.* **8**, 2241–2253.
Scherer, S. and Davis, R. W. (1980). *Science* **209**, 1380–1384.
Schimke, R. T. (1980). *Sci. Am.* **243**, 50–59.
Sidloi, R., Kleiman, L. and Schulman, H. M. (1978). *Nature, Lond.* **273**, 558–560.
Slightom, J. L., Blechl, A. E. and Smithies, O. (1980). *Cell* **21**, 627–638.
Smith, D. F., McClelland, A., White, B. N., Addison, C. F. and Glover, D. M. (1981). *Cell* **23**, 441–449.
Smith, T. F. (1980). *Nature, Lond.* **285**, 620.
Southern, E. M. (1975). *J. Mol. Biol.* **98**, 503–517.
Southern, E. M. (1980). *In* "Methods in Enzymology" (Ed. R. Wu) Vol. 68, 152–176. Academic Press, New York.
Spradling, A. C., Digan, M. E., Mahowald, A. P., Scott, M. and Craig, E. A. (1980). *Cell* **19**, 905–911.
Spritz, R. A., DeRiel, J. K., Forget, B. G. and Weissman, S. M. (1980). *Cell* **21**, 639–646.
Stein, J. P., Catterall, J. F., Kristo, P., Means, A. R. and O'Malley, B. W. (1980). *Cell* **21**, 681–687.
Streuli, M., Nagata, S. and Weissmann, C. (1980). *Science* **209**, 1343–1347.
Sullivan, D., Brisson, N., Goodchild, B., Verma, D. P. S. and Thomas, D. Y. (1981). *Nature, Lond.* **289**, 516–518.
Sun, S. M., Slightom, J. L. and Hall, T. C. (1981). *Nature, Lond.* **289**, 37–41.
Szostak, J. W. and Wu, R. (1980). *Nature, Lond.* **284**, 426–430.
Takahashi, N., Nakai, S. and Honjo, T. (1980). *Nucl. Acids Res.* **8**, 5983–5991.
Taniguchi, T., Mantei, N., Schwartzstein, M., Nagata, S., Muramatsu, M. and Weissmann, C. (1980). *Nature, Lond.* **285**, 547–549.
Tartof, K. D. (1975). *A. Rev. Genet.* **9**, 355–365.
Temin, H. M. (1980). *Cell* **21**, 599–600.
Tobin, S. L., Zulaf, E., Sanchez, F., Craig, E. A. and McCarthy, B. J. (1980). *Cell* **19**, 121–131.
Tonegawa, S., Maxam, A. M., Tizard, R., Bernard, O. and Gilbert, W. (1978). *Proc. Natn. Acad. Sci. U.S.A.* **74**, 3171–3175.

Tuan, D., Biro, P. A., deRiel, J. K., Lazarus, H. and Forget, B. G. (1979). *Nucl. Acids Res.* **6**, 2519–2544.
Upholt, W. B. (1977). *Nucl. Acids Res.* **4**, 1257–1265.
Van Den Berg, J., van Ooyen, A., Mantei, N., Schamböck, A., Grosveld, G., Flavell, R. A. and Weissmann, C. (1978). *Nature, Lond.* **276**, 37–44.
Van Der Ploeg, L. H. T. and Flavell, R. A. (1980). *Cell* **19**, 947–958.
Van Der Ploeg, L. H. T., Konings, A., Oort, M., Roos, D., Bernini, L. and Flavell, R. A. (1980). *Nature, Lond.* **283**, 637–642.
Van Ooyen, A., Van Den Berg, J., Mantei, N. and Weissmann, C. (1979). *Science* **206**, 337–344.
Vandekerckhove, J. and Weber, K. (1980). *Nature, Lond.* **284**, 475–477.
Vanin, E. F., Goldberg, G. I., Tucker, P. W. and Smithies, O. (1980). *Nature, Lond.* **286**, 222–226.
Vogeli, G., Avvedimento, E. V., Sullivan, M., Maizel, J. V., Lozano, G., Adams, S. L., Pastan, I. and de Crombrugghe, B. (1980). *Nucl. Acids Res.* **8**, 1823–1837.
Wahli, W., Dawid, I. B., Wyler, T., Jaggi, R. B., Weber, R. and Ryffel, G. U. (1979). *Cell* **16**, 535–549.
Wahli, W., Dawid, I. B., Wyler, T., Weber, R. and Ryffel, G. U. (1980). *Cell* **20**, 107–117.
Wallace, R. B., Johnson, P. F., Tanaka, S., Schöld, M., Itakura, K. and Abelson, J. (1980). *Science* **209**, 1396–1400.
Weldon Jones, C. and Kafatos, F. C. (1980). *Nature, Lond.* **284**, 635.
Williams, L. A. and Piatigorsky, J. (1979). *Eur. J. Biochem.* **100**, 349–357.
Wilson, A. C., Carlson, S. S. and White, T. J. (1977). *A. Rev. Biochem.* **46**, 573–639.
Woese, C. R. and Fox, G. E. (1977). *Proc. Natn. Acad. Sci. U.S.A.* **74**, 5088–5090.
Zimmer, E. A., Martin, S. L., Beverley, S. M., Kan, Y. W. and Wilson, A. C. (1980). *Proc. Natn. Acad. Sci. U.S.A.* **77**, 2158–2162.

The use of genomic libraries for the isolation and study of eukaryotic genes

H. H. DAHL, R. A. FLAVELL and F. G. GROSVELD

Laboratory of Gene Structure and Expression, National Institute for Medical Research, Mill Hill, London, UK

I Introduction

In nearly all organisms, DNA is the carrier of genetic information (review, Olby, 1974). One of the main questions in modern molecular biology has been: how is this genetic information organized and how is its expression regulated? Until recently methods for analysing gene structure and expression, particularly in eukaryotes, were few and slow to yield results, and our knowledge, therefore, was restricted. This was mainly due to the fact that a single gene normally is only a very small proportion of the genomic DNA.

The DNA content per haploid genome differs significantly between various organisms. A bacterium such as *E. coli* has the potential to code for 3000—4000 different gene products. In humans the corresponding number is approximately 10^6, although only a small part of this DNA encodes proteins. Not all eukaryotic DNA is transcribed and one expects to find regulatory sequences, perhaps with (nonfunctional?) spacer sequences, interspersed in the DNA. Classical studies of inheritance, hybridization analysis, electron microscopy, and fine structure gene mapping have provided some insight into gene organization, especially in prokaryotic cells (Benzer, 1962). However, it is only with the introduction of recombinant DNA technology in the late 1960s and 1970s that we have been able to isolate specific single copy genes or DNA fragments, purify them away from other sequences and analyse them in close detail. This new technology was made possible by:

- the discovery of restriction endonucleases which cleave DNA at specific nucleotide sequences.
- the development of prokaryotic and eukaryotic cloning vehicles (vectors), which are replicons into which foreign DNA can be inserted. Most suitable in prokaryotes are small plasmids and bacteriophages.
- development of chemical and biological methods whereby DNA fragments from various sources can be covalently linked.
- development of techniques for reintroduction of the recombinant DNA into cells with high efficiency.
- screening methods that allow selection for specific recombinant DNA sequences.

In view of the importance of genetic engineering and molecular cloning, it is not surprising that development in these fields has been both fast and extensive. These advances have made it possible to construct complete gene libraries (or gene banks) from any organism. This involves generation of a complete collection of cloned DNA fragments, which comprise the entire genome of the organism. The clones are produced by the following steps:

- making clonable DNA fragments that in sum contain the total genome of the donor organism;
- joining these fragments to a suitable cloning vector;
- introduction of the recombinant molecules into host cells at high efficiency to obtain a large number of independent clones. If this number is large enough, then the entire genome will be represented in the library.

Aspects of making genomic libraries and their screening for specific cloned gene sequences are reviewed below.

Recombinant DNA technology provides us with powerful methods for purification of specific DNA sequences from other host cell DNA sequences and has been of singular interest in three different areas, namely:

- isolation of a specific and identified DNA sequence in large amounts. This makes it possible to analyse DNA structure in great detail by restriction enzyme analysis and DNA sequencing. By comparing the primary structure of a protein or messenger RNA with that of the genomic DNA sequence, analysis of coding sequences (and intervening sequences in eukaryotic cells) can be undertaken. By comparison of the structure of different genes (for example those regulated co-ordinately) putative sequences involved in expression or regulation of a gene may be located and defined, and if test systems are available to measure gene expression to ask if a given sequence is required for that expression.
- analysis of gene expression either *in vitro*, or by reintroducing the gene(s) as recombinant DNA into cells. By combining these approaches with *in vitro* mutagenesis of recombinant DNA, we may gain useful information about DNA regions important for controlling gene expression.
- expression of proteins, hormones or other biological molecules important for science, medicine or industry. These molecules are at present often produced at high cost and in a relatively impure state. Examples are various human hormones like insulin, growth hormone and the antiviral agent, interferon. By cloning the genes coding for these products in a form that allows expression of the cloned DNA to give a biologically active molecule, a large quantity of the product can be isolated at low cost and in a system that affords high purity.

Before starting a cloning experiment, it is necessary to consider what type of clones are wanted:

- cDNA or genomic DNA clones;
- clones from DNA fragments where the frequency of the desired gene segment has been increased by an enrichment procedure, or from total gene libraries.

cDNA clones (Williams, this series, Vol. 1) are constructed by synthesizing complementary cDNA from an RNA. Since poly(A)$^+$-selected mRNA is the usual starting material, the cDNA clones will reflect the structure of mRNA and therefore contain no intervening sequences. Furthermore, the distribution of clones will also usually reflect that of the mRNA, which makes isolation of cDNA clones corresponding to an abundant mRNA relatively easy, but more difficult when the mRNA is present in low amounts. Finally, non-transcribed regions cannot be cloned this way.

When genomic DNA is used as starting material, a choice has to be made between making a total gene library, or enriching for the desired gene fragment(s). Preliminary purification of the DNA segments reduces the total number of clones that need to be screened. This is an advantage when the isolation of a specific gene fragment is wanted and the restriction enzyme sites around the gene are known. It is particularly suitable for the isolation of the same DNA segment from a large number of different DNA samples. Isolation of clones from a total genomic library requires screening of a larger number of clones than required when purified DNA fragments are used, but it gives a larger flexibility, since all DNA sequences should be present and overlapping DNA fragments can be isolated to facilitate mapping.

II Cloning of gene libraries

The general strategy for the construction of a gene library first requires the isolation of DNA fragments from the donor organism. These fragments must, of course, in sum contain the entire genome, and can be made from total genomic DNA by shearing or by enzymatic cleavage. The genomic DNA fragments are then joined to a suitable cloning vehicle (vector) *in vitro*. Usually the vector and donor DNA fragments are covalently joined by use of the enzyme DNA ligase. These recombinant DNA molecules are introduced into cells (usually *E. coli*), where they must replicate, and usually also express at least the genes coded for by the vector DNA to permit growth and selection. The desired donor DNA fragment can then be identified and analysed in detail (Fig. 1).

It is axiomatic in the cloning procedure that sufficient clones are obtained to contain the entire genome. The precise number of clones that is required depends on both the size of the donor genome and the average size of the DNA fragments cloned. The most common vector systems are bacteriophage λ, plasmids and cosmids (see Section III), and the cloning capacities of these vectors are usually up to a

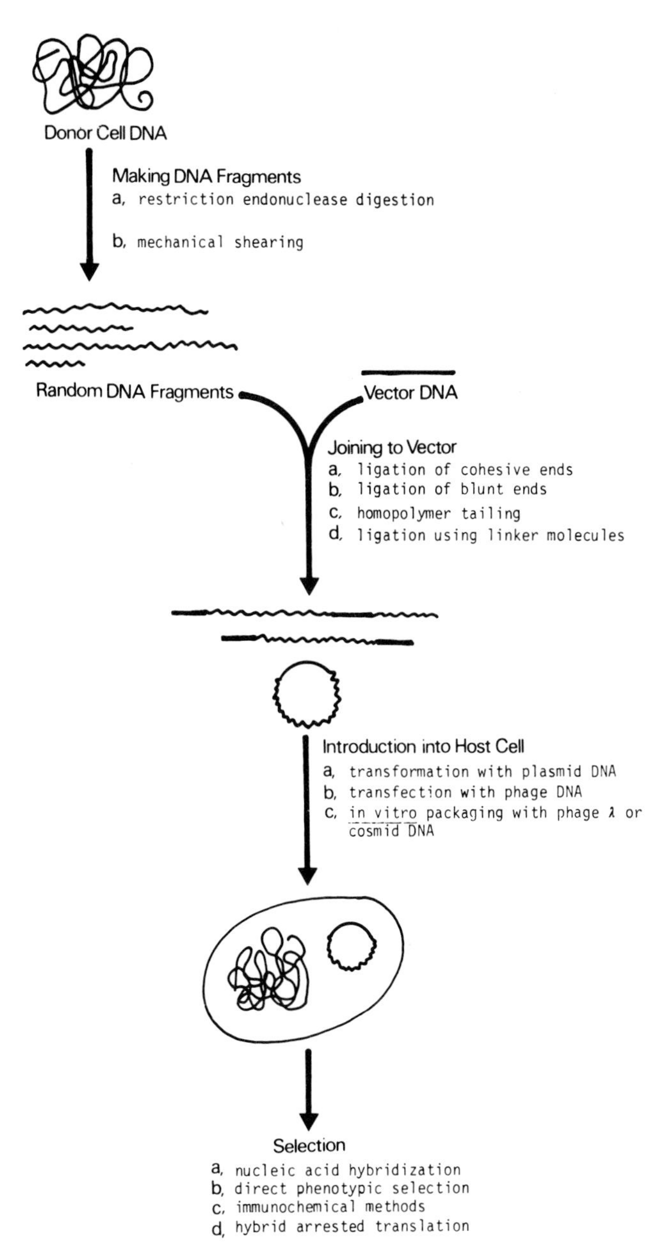

Figure 1 Schematic illustration of the procedure for construction of gene libraries and isolation of desired clones. DNA from a donor organism is degraded to give random fragments. These fragments are then joined to a vector (e.g. phage λ, plasmid or cosmid DNA). Plasmids, cosmids and often also phage λ vectors have selective markers, so that bacteria containing recombinant DNA can be recognized. Finally, the bacteria containing the desired gene(s) are selected for and isolated. a, b etc. represent alternative procedures for each respective step.

maximum insert size of approximately 25 kb, 15 kb and 45 kb (kilobase pairs) respectively. Table 1 lists the theoretical number of clones required to represent the total genome from various organisms. However, because DNA fragments are cloned on a random basis, this is the minimum number of clones required for a genomic library. Statistically, this means that when screening a library containing the minimum number of recombinant DNA clones, there is a 50% chance of finding a given single copy gene. When screening twice the minimum number of clones, the chance of finding a given single copy gene is 75%, and so on. It is therefore necessary to screen 3—10 times this minimal number of clones, in order to have a reasonable chance of identifying a particular gene.

Table 1 The number of clones theoretically required to represent the genome once from various organisms.*

Size of cloned fragments (bp)	Genome size (bp)		
	2×10^6 (bacteria)	2×10^7 (fungi)	3×10^9 (mammals)
5×10^3	400	4000	600 000
10×10^3	200	2000	300 000
20×10^3	100	1000	150 000
40×10^3	50	500	75 000

*Numbers are derived by dividing the genome size of the donor organisms (in this case, the approximate number of base pairs from bacteria, fungi and mammals, respectively) with the average size of the cloned fragments.

Construction of total gene libraries makes it relatively easy to isolate any DNA fragment (gene) for which a probe is available (see Section VI). Furthermore, random, overlapping DNA fragments can be used in making total gene libraries, which has advantages when:

- the DNA region to be studied is too large to be contained in one recombinant clone (as for the human β-globin region); the availability of overlapping DNA fragments makes possible the rapid determination of the gene organization in the region of interest.
- "walking" along the genome is planned.

A DNA cloning

The DNA to be cloned can be inserted in the vector in a number of different ways. The decision as to which one to use depends on the specific situation, but it is of importance that the DNA is of a purity and size that permits enzymatic manipulation.

When the starting material is double stranded genomic DNA, several possibilities exist for the preparation of the donor DNA. Most straightforward is to digest the DNA with a restriction enzyme. If the physical map in and around the required DNA fragment is known, a complete digest with one or two restriction enzymes might give clonable DNA of a suitable size and containing the desired sequences. When double digest DNAs are cloned, it is possible, by digesting the vector with the same enzymes, or enzymes giving the same overlapping ends, to direct the orientation of the cloned fragments with respect to the vector.

When constructing a gene library (or when the physical map of restriction enzyme sites around the desired gene is not known) DNA fragments that together comprise the entire genome are required. Although it is possible to imagine these DNA fragments being made in a specific, rather than a random, way — for example, as a total restriction enzyme digest — the risk exists that some fragments are either too large to be cloned or too small to be detected in the screening procedure. Random DNA fragments are therefore by far the most useful starting material for gene libraries, and these random DNA fragments can be created by mechanical shearing (Wensink *et al*., 1974; Clarke and Carbon, 1976), or by partial digestion with one (or more) restriction endonucleases (Sinsheimer, 1977). It should be noted, however, that we do not know whether the DNA fragments created by either of these techniques are totally random.

When choosing restriction enzyme(s) to generate DNA fragments, it is important to choose enzymes that optimize the chance that every sequence is ultimately represented in the library (Maniatis *et al*., 1978). Although total and partial Eco RI digests of DNA, for example, have successfully been used for cloning several eukaryotic genes (e.g. Fritsch *et al*., 1980), this λ library will not contain DNA regions where Eco RI sites are more than 25 kb apart. This risk can be minimized by choosing restriction enzymes that recognize 4 bp (base pair) sequences and, therefore, usually cut the DNA more often than enzymes recognizing 6 bp sequences. If we assume that the four deoxyribonucleotides in DNA are present in equal amounts and randomly distributed, a particular tetranucleotide sequence should occur once in every 256 (4^4) bp. Similarly, a given hexanucleotide sequence should occur once in 4096 (4^6) bp. The aim is to generate random DNA fragments of a given size, in the case of a λ vector usually approximately 20 kb. The number of ways to generate 20 kb fragments increases with the number of possible cleavage sites. Partial digestion with restriction enzymes that recognize a 4 bp sequence will therefore result in a more random collection of DNA fragments as against fragments obtained by digestion using enzymes recognizing 6 bp sequences. Examples of enzymes with a four base pair recognition sequence that have been used in the construction of gene libraries are Hae III, Alu I and the isoschizomers Mbo I and Sau 3A.

However, experience has shown that libraries constructed in this way lack certain sequences. For example, T. Maniatis and co-workers constructed a human DNA library by cloning DNA fragments obtained by partial digestion with restriction enzymes Hae III and Alu I (Lawn *et al.*, 1978). No clones containing the γ-globin region or the sequence between the γ- and δ-globin genes were found, although clones containing other regions of the β-globin gene locus were isolated. The γ-globin gene region was isolated by making a library from human DNA that was partially digested by Eco RI (Maniatis *et al.*, 1978). In this case, no clones containing the region between the γ- and δ-globin genes were found. This region was eventually isolated by cloning a partially purified fragment (Fritsch *et al.*, 1980), although we have isolated several such clones containing this DNA region from human gene libraries constructed from partial Mbo I-digested DNA (Van Der Ploeg *et al.*, 1980; J. Groffen, unpublished results). The reason why certain regions are not present, or underrepresented, in some enzyme digests and not others is not known. One possibility is that certain restriction enzyme sites are preferentially cut and that the resulting DNA fragments, therefore, are not randomly distributed. Also, the libraries constructed by T. Maniatis and co-workers were amplified, and the amplification step might lead to selective loss of specific sequences, due to slow replication, production of toxic substances by certain clones, or by further recombination of the recombinant DNA molecule.

B Size fractionation of DNA

When constructing a gene *library*, in general it is not desirable to enrich for a specific gene, since this will eliminate other DNA sequences which may be required in the future. It may, however, be an advantage to size fractionate the donor DNA in the following situations:

- to clone DNA fragments of sizes that approach the maximum cloning capacity of the vector used. This minimizes the risk that several small donor DNA fragments will be inserted in the same vector, linking non-contiguous segments of the genome in a single clone, which is obviously undesirable.
- the DNA to be cloned is known to be within a predetermined size range.
- DNA fragments of a size not suitable for cloning will, in

unfractionated DNA preparations, be ligated to vector molecules and will lower the efficiency of introduction of the recombinant DNA into cells (as noticed when the λ phage packaging system is used, see below); small fragments are removed by the size fractionation.

Size fractionation of DNA is usually by agarose gel electrophoresis (Aaij and Borst, 1972) or sucrose gradient centrifugation. By agarose gel electrophoresis it is easy to obtain a DNA fragment of high purity, or DNA which is homogeneous in size, within a very narrow range. Sucrose gradient centrifugation is also a fast method for obtaining DNA fragments, although size separation is usually not as good as with agarose gel electrophoresis. However, agarose gel electrophoresis of fractionated DNA can give problems in practice, because contaminants from the agarose sometimes inhibit subsequent enzymatic reactions.

C Gene enrichment

If the aim of cloning experiments is to isolate a *specific gene*, it will make the cloning procedure simpler and faster if the desired gene fragment can be enriched. At the moment, relatively few methods are available for fractionation of DNA, due to the general lack of sequence-related physical differences, and most methods are based on the size fractionation procedures mentioned above.

A method that fractionates DNA according to other properties as well as size is reverse phase chromatography, using the matrix RPC-5 (Hardies and Wells, 1976). The basis for separation of DNA fragments using RPC-5 chromatography is not known, but the method gives high recovery of clonable DNA and commonly gives 10-fold enrichment of a specific sequence (Tilghman *et al.*, 1977; Edgell *et al.*, 1979). When chromosomal DNA is fractionated, it is essential that the elution of the desired gene can be monitored, as by Southern blotting (Southern, 1975), since there is no *a priori* means of predicting the elution profile of a given DNA sequence. Quantitative Southern blotting can also be used to estimate the enrichment obtained.

It also follows that to use these methods for gene enrichment, the desired gene must be located on identical DNA fragments, rather than an overlapping collection of fragments. The DNA to be cloned cannot therefore be random DNA fragments, but is normally a total restriction enzyme digest of the donor DNA.

III Vector systems

A Plasmids

Plasmids are extra-chromosomal genetic elements and are associated with sex factors, drug resistance, colicin factors and phages (Meynell,

1972). They exist as free supercoiled circular DNA in the bacterium and are therefore easy to purify. Since the first report on cloning DNA using plasmids as vectors (Cohen *et al.*, 1973), a large number of plasmids — especially for use in cloning using *E. coli* as host — have been constructed (see Morrow, 1979). When used in recombinant DNA work, it is desirable that the plasmid has the following features:

— the plasmid must be capable of autonomous replication
— because the most common cloning techniques involve the use of specific restriction endonucleases, the plasmid should have an appropriate spectrum of cleavage sites, in general with single cleavage sites for as many restriction enzymes as possible
— the restriction enzyme sites used for insertion of the donor DNA should be positioned so that a plasmid which contains an insert still expresses a physical marker useful for selection
— since only a small proportion of cells are transformed with the recombinant DNA in a cloning experiment, it is virtually essential that the plasmids confer a selective phenotype on the transformed host cell, such as drug resistance or immunity to specific colicins.

Most plasmids used for cloning in *E. coli* have been derived from the plasmids pSC101 or Col E1 by deletion, insertion or mutation (Sinsheimer, 1977; Curtiss, 1976). pSC101 (Cohen *et al.*, 1973) transfers tetracycline resistance, while Col E1 (Hershfield *et al.*, 1974) confers immunity to colicin E1. Both these plasmids replicate autonomously in the bacterium. However, pSC101 (like chromosomal DNA) replicates in a "stringent" fashion (Timmis *et al.*, 1974), which means that it requires protein synthesis for initiation of replication, but not active DNA polymerase I. Col E1 replicates in a "relaxed" fashion (Clewell, 1974), which means that it is independent of protein synthesis; it requires active DNA polymerase I. These plasmids are present in a copy number of 6—8 and 20—30 plasmids/cell, respectively, but in cells containing Col E1 (and many of its derivatives), the copy number can be amplified to several hundred — up to 50% of the total cellular DNA — after administration of the protein synthesis inhibitors chloramphenicol, tetracycline or spectinomycin to exponentially growing cells. These block chromosomal, but not plasmid, DNA replication (Clewell, 1974). Although it is usually of interest to have as high a copy number as possible, situations might arise where a low copy number is desirable, as in cases where cloned gene products in large amounts are lethal to the host cell.

To insert a foreign piece of DNA in the plasmid, the latter is usually cleaved with one or more restriction endonucleases. The restriction enzyme map of the plasmid is therefore of importance and it is usually desirable that the plasmid has only one cleavage site for the restriction enzyme used. For example, pSC101 has single Eco RI, Bam HI, Hind III and Sal I sites and Col E1 has single Eco RI and Sma I/Xma I sites. Partly because small DNA molecules give higher transformation efficiencies compared to larger DNA molecules when using the $CaCl_2$ transformation method (see Section V.A), and partly because they are easier to manipulate, the trend in plasmid-vector design has been to make relatively small plasmids (containing 3—10 kb of DNA) with a large number of single restriction enzyme cleavage sites. For example, the much used plasmid pBR322 (Bolivar *et al.*, 1977a, b) contains the tetracycline resistance gene from pSC101 and the ampicillin resistance gene derived from the TnA segment of an R-factor, together with a part of Col E1 which contains the replication origin among other sequences. The resulting plasmid pBR322 confers ampicillin and tetracycline resistance to the bacterium, and it has single cloning sites for the restriction enzymes Eco RI, Hind III, Sal I, Bam HI and Pst I. Like Col E1 it replicates in a relaxed fashion.

A large number of plasmid cloning vectors have been constructed, some of which are mentioned in Table 2. The plasmids pCRI and pMB9 (Armstrong *et al.*, 1977; Bolivar *et al.*, 1977a) are both derived from Col E1 and apart from transferring colicin EI immunity, they also confer kanamycin and tetracycline resistance respectively.

Another requirement is that the plasmid, in combination with the host and the physical containment facilities, should provide maximum safety with respect to potential biological hazards. To reduce the risk of a recombinant plasmid establishing itself in bacteria, for instance in the human gut where the cloned gene might be expressed and be harmful to the host, the plasmids are designed not to be self-transmissible. Usually they cannot be mobilized via conjugative plasmids. An example is pBR322 which can only be mobilized at a very low, but measurable, frequency. However, Twigg and Sherratt (1980) removed an Hae II fragment from pBR322 and the resulting plasmid, pAT153, cannot be mobilized by conjugated transfer. pAT153 is now widely used in the UK as a "safe" vector.

Insertion of DNA into restriction enzyme sites in plasmids leads to inactivation of a genetic marker, if the marker gene sequence is interrupted. If DNA is inserted in the Hind III, Sal I or Bam HI sites of pBR322 (or pAT153), the result will be loss of tetracycline resistance (although, since the Hind III site is in the promoter region of the tet gene, certain DNA inserts reconstitute the promoter and the plasmids are therefore tetracycline resistant), and bacteria containing plasmids with insertions can be selected for by their ability to grow in the presence of ampicillin but not tetracycline (Bolivar, 1978). In practice not all tetracycline-sensitive bacteria contain plasmids with inserts (Bolivar, 1978); some are mutants caused by

Table 2 **Characterization of some plasmid cloning vectors.**

Plasmid	Useful cloning sites	Selective markers*	Sites at which insert inactivates marker	Molecular weight (kb)	Amplifiable†	Reference
Col E1	Eco RI, Sma I/Xma I Hind II, Pst I (2 sites)	El imm	Colicin production: Eco RI, Sma I/Xma I, Pst I	7.0	+	Hershfield *et al.* (1974)
pSC101	Eco RI	Tet^R	Tet: none useful	9.7	-	Cohen *et al.* (1973)
pCR1	Eco RI, Hind III, Sal I	Kan^R, El imm	Kan: Hind III Colicin production: Eco RI	14.8	+	Covey *et al.* (1976)
pMB9	Eco RI, Hind III, Sal I, Bam HI, Sph I	Tet^R, El imm	Tet: Hind III, Sal I, Bam HI, Sph I Colicin production: Eco RI	5.9	+	Rodriguez *et al.* (1976)
pBR322	Eco RI, Hind III, Sal I, Bam HI, Pst I, Pvu I, Pvu II, Xma III, Sph I	Amp^R, Tet^R	Amp: Pst I, Pvu I Tet: Hind III, Sal I, Bam HI Sph I	4.4	+	Bolivar *et al.* (1977b)
pBR328	Eco RI, Hind III, Sal I, Bal I, Pvu II, Bam HI, Pst I, Pvu I, Hind II (2 sites), Sph I	Cm^R, Amp^R, Tet^R	Cm: Eco RI, Bal I, Pvu II Amp: Pst I, Hinc II, Pvu I Tet: Hind III, Sal I, Bam HI, Hinc II Sph I	4.9	+	Soberon *et al.* (1980)
pAT153	Eco RI, Hind III, Sal I, Xma III, Bam HI, Pst 1, Pvu I, Sph I	Amp^R, Tet^R	Amp: Pst I, Pvu I Tet: Hind III, Sal I, Bam HI, Sph I	3.7	+	Twigg and Sherratt (1980)
pBR327	Eco RI, Hind III, Sal I, Bam HI, Pst I, Pvu I, Sph I	Amp^R, Tet^R	Amp: Pst I, Pvu I Tet: Hind III, Sal I, Bam HI, Sph I	3.3	+	Soberon *et al.* (1980)

*Ampicillin resistance (Amp^R); colicin production (EI imm); tetracycline resistance (Tet^R); kanamycin resistance (Kan^R); chloramphenicol resistance (Cm^R).
†+ = copy number of the plasmid is increased to several hundred in the presence of chloramphenicol.

minor base changes at the religated restriction enzyme site. Although it often makes the initial screening of recombinants easier, inactivation of physical markers is not essential to the cloning procedure.

Plasmids have successfully been used for construction of gene libraries from *Xenopus laevis* (Smith *et al.*, 1979), *Drosophila melanogaster* (Glover *et al.*, 1975), sea urchin (Kedes *et al.*, 1975) and yeast (Carbon *et al.*, 1977). Due to the decreased efficiency of transformation of large plasmids (see Section V.A), recombinants with inserts longer than 10–15 kb are seldom detected and the average size of inserts is usually 5–10 kb. Colony hybridization methods have been developed which make the screening for specific clones relatively easy (see Section VI). It is also easy to purify large amounts of plasmid from a clone, because every bacterium contains the recombinant plasmid, bacteria multiply quickly, and rapid methods for isolating pure plasmids have been developed (Kahn *et al.*, 1979).

B Bacteriophage λ vectors

Coliphage λ has been extensively studied, and recently λ mutants for use as cloning vehicles have been constructed. The genome of wild-type bacteriophage λ contains a double-stranded DNA of approximately 49 000 bp (Schroeder and Blattner, 1978). Following infection of *E. coli*, the phage can replicate in a lytic or lysogenic cycle (Fig. 2; Hershey, 1971). If phage λ is used as a cloning vehicle, the phage must replicate in a lytic cycle where, following infection, the λ DNA cohesive ends (cos) are joined and the double-stranded DNA circle replicates. Proteins required for the assembly of the λ phage particle and lysis of the host cell are synthesized late in infection. Finally, the DNA will be packaged into a phage particle when two cos sites — in the same orientation — are separated by from 35–53 kb of DNA (for a review see Earnshaw and Casjens, 1980). Although phage λ has about 50 genes, only about 50% of these are necessary for growth and plaque formation (Fig. 3). Non-essential DNA sequences can therefore be replaced by donor DNA and the inserted DNA propagated as a phage λ recombinant. The λ cloning system has the following advantages when compared with plasmids:

- the DNA can be packaged *in vitro* into λ phage particles and transduced into *E. coli* with high efficiency (see section on *in vitro* packaging of λ phage).
- foreign DNA up to 25 kb in length can be inserted into a λ phage vector (Williams and Blattner, 1979).

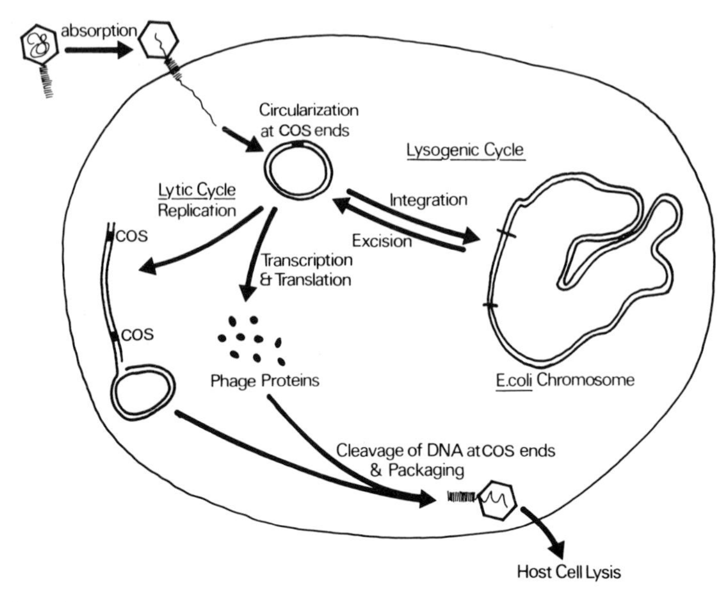

Figure 2 Illustration of phage λ lytic and lysogenic cycles. After infection of *E. coli* the phage λ DNA circularizes. In the lysogenic cycle this DNA integrates into the chromosomal DNA, where it replicates as a chromosomal DNA segment; no phage structural proteins are synthesized. In the lytic cycle phage DNA replication occurs to yield, ultimately, several hundred phage particles per *E. coli* cell.

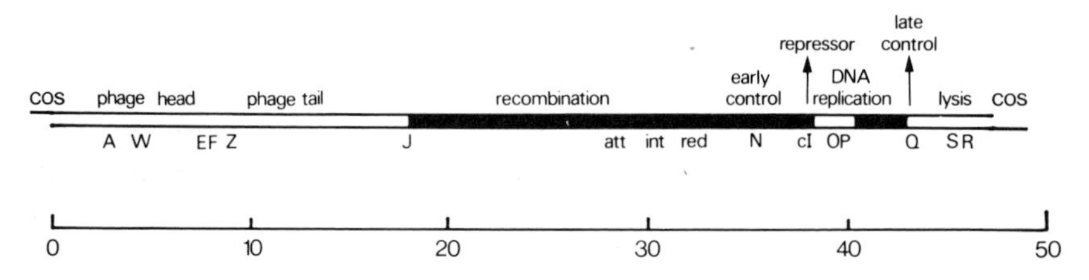

Figure 3 Essential and non-essential regions in the phage λ genome. The black segments indicate regions that are non-essential for plaque formation and phage growth. A number of regions coding for various phage functions are indicated, as are the cohesive ends (cos) and genes mentioned in the text. The size of phage λ DNA is indicated in kilobase pairs (kb).

— screening and storage of the λ recombinant DNA is easier (Benton and Davis, 1977).

Before λ DNA could be used effectively as a cloning vehicle, two problems had to be solved. The first was to create λ mutants with suitably placed restriction enzyme sites, and into which foreign DNA could easily be inserted without affecting essential λ phage functions. Since wild-type λ DNA is large, it has several sites for most restriction enzymes used for cloning (e.g. five for Eco RI and six for Hind III). Many of these sites had to be eliminated by mutation or deletion before a useful cloning vector could be obtained.

The second problem was to construct λ vectors in which large DNA fragments could be inserted. There is the upper size limit for DNA that can be packaged into viable λ particles; thus the maximum size of a foreign DNA fragment that can be inserted into wild-type λ DNA is approximately 5 kb.

Two types of phage λ cloning vectors have been constructed: insertion and replacement vectors. Insertion vectors have only one cleavage site for the restriction enzyme used and insertion of DNA at this site must not affect essential λ phage functions. Since these λ phages also have to be propagated as non-recombinant phages, and because there is both an upper and lower limit of the size of DNA that can be packaged into a phage particle (approximately 35—53 kb), the minimum size of the vector alone must be about 35 kb, which means that the maximum size of an insert is approximately 18 kb.

In replacement vectors the maximum size of the insert DNA fragment depends on how much of the λ DNA is non-essential (and therefore can be substituted), and on the upper DNA size limit for packaging into phage particles. About 25—30 kb of the λ genome codes for gene products essential for the lytic phage cycle. It therefore follows that maximum sizes of foreign DNA in replacement λ vectors is approximately 20—25 kb. If one wants to replace one fragment with another, the restriction enzyme sites used should occur at least twice and the deleted part must be non-essential for the λ lytic cycle. A number of λ phages have now been constructed which fulfil these requirements (Morrow, 1979). Some of these phages are listed in Table 3. The λ cloning vectors are usually made from λ mutants obtained by deleting parts of the genome. Substitution of parts of related phage genomes (for example Ø80) into λ also proved useful in designing vectors for cloning, not only to provide altered restriction site maps, but also because it permitted convenient rearrangement of essential genes on the genomic λ DNA.

An example of a substitution vector is λgt WES. The λgt WES phages are derived from λgt (Tiemeier *et al.*, 1976). This was done by eliminating two Eco RI sites and by introducing two deletions in the central non-essential part of the λ DNA. The resulting phages were called λgt · λB and λgt · λC. These two phage DNAs each contain two Eco RI sites, and when digested with this enzyme three fragments are produced: two large outer fragments containing the essential genes and a central piece (either the Eco RI B or C fragment) which contains non-essential genes. The central fragment can therefore be substituted by a foreign DNA fragment (1—14 kb in

Table 3 Characterization of some phage λ cloning vectors.

Phage	Useful cloning sites	Markers for hybrid screening	Cloning capacity (kb)	Reference
λgt WES	Eco RI, Sst I	None	2.4–17.4	Tiemeier *et al.* (1976)
Charon 3	Eco RI	None	0– 8.6	De Wet *et al.* (1980)
Charon 4	Eco RI	lac^-, bio^-*	7.3–19.3	De Wet *et al.* (1980)
Charon 21A	Eco RI, Hind III, Xho I, Sal I	None	0– 8.2	De Wet *et al.* (1980)
λ1059	Bam HI	spi^-†	6.3–24.4	Karn *et al.* (1980)
NM 641	Eco RI	Clear plaques	0–12.5	Murray *et al.* (1977)
NM 762	Hind III	Colourless plaques‡	3.4–18.4	Murray *et al.* (1977)

*lac^- gives colourless plaques on lac^+ cells; bio^+ phenotype is detected by growth of bacteria around plaques of bio^- cells on biotin deficient plates.

†spi^- phenotype is detected by its ability to grow on a bacterial strain lysogenic for phage P2.

‡Colourless plaques are obtained when recombinant phages are plated on a host carrying an amber mutation in the lac 2 gene.

length). The λgt system has a further advantage: although the two outer DNA fragments provide all the necessary information for phage propagation, their combined length is not sufficient for phage packaging. It is therefore essential to have a DNA insert in order to produce viable phages, and this provides a selection system for recombinants. Only λ DNA with inserts will be able to replicate and form plaques when plated on *E. coli*.

In order to contain the λ recombinant phage in laboratory bacterial strains (biological containment) three amber mutations have been introduced in λgt WES: genes *W* (necessary for phage assembly), *E* (necessary for coat protein synthesis) and *S* (necessary for host cell lysis). The resulting λ phages, λgt WES · λB and λgt WES · λC, are therefore only able to grow on *E. coli* hosts that suppress the first two mutations (suII), and only lyse hosts (suIII) that suppress the amS mutation.

Blattner and colleagues have constructed a number of λ cloning vectors of both the insertion and substitution type (Williams and Blattner, 1979). So far, 21 of these so-called Charon phages have been described (De Wet *et al.*, 1980).

A number of these phages contain the lac 5 region from *E. coli* which codes for β-galactosidase and in which an Eco RI site is situated in the distal region of the lac Z promotor. This means that these Charon phages, when plated on the colourless chromogenic substrate 5-bromo-4-chloro-3-indolyl-β-D-galactoside (XG), give blue plaques, due to synthesis of β-galactosidase, which cleaves the chromogenic substrate and releases a blue indolyl derivative. In recombinant phages the central Eco RI fragment containing the lac Z promotor is substituted and β-galactosidase is therefore not produced by the phage, which therefore gives colourless or faint blue plaques (Williams and Blattner, 1979). The maximum cloning capacity in these Charon phages is 22 kb (De Wet *et al.*, 1980).

Plaque turbidity has also been useful when selecting recombinant λ phages. Plaque turbidity is determined by the presence of non-lysed bacteria. When the repressor coded for by the immunity region gene CI is expressed, some infected cells will enter the lysogenic cycle and plaques produced will therefore be turbid. When DNA is inserted into this gene, a functional repressor cannot be made and all bacteria are lysed. As a consequence, these recombinant phages give clear plaques, which are easily distinguished from the turbid plaque morphology of the parental λ phage synthesizing the repressor. For example, a number of λ phage vectors constructed by N. Murray *et al.* contain the CI gene from imm 434 (Murray *et al.*, 1977). This gene contains Eco RI and Hind III sites, which can be used when inserting a foreign DNA fragment and selecting recombinants by the absence of plaque turbidity.

Recently a novel type of λ cloning vector has been described (Karn *et al.* 1980); λ 1059 is a substitution vector composed of three Bam fragments. All the essential λ functions are contained on the two arms of the vector. The central part contains the λ red and gamma genes, which are under the control of the promoter pL and λ repressor (c1857). These genes confer the spi$^+$ phenotype on the vector and it is there able to grow on rec A host (e.g. *E. coli* Q358), but not on strains lysogenic for phage P2 (e.g. *E. coli* Q389) (Lindahl *et al.*, 1970). When the central Bam fragment is removed, ligation of the two vector arms does not result in viable phages because their length is too short to fill the phage head properly. Viable phages are produced when the central Bam fragment is substituted with a 6—24 kb foreign DNA insert. These recombinant phages have the spi$^-$ phenotype and are now able to grow on the P2 lysogen strain. Recombinants are therefore easily constructed by digesting the λ 1059 DNA with Bam HI, joining it to donor DNA of suitable size, and which has been cleaved with a restriction enzyme that gives the same overlapping ends as Bam HI (e.g. Bam HI, Bgl II, Sau 3A, Mbo I) and plating the mixture on *E. coli* strain Q389. Only recombinant phages will produce plaques.

Phage λ DNA has been, until now, the most widely used phage cloning vehicle, because phage λ cloning methods were developed early and because it is relatively easy to create, handle and screen a large number of recombinant DNA-containing phages. The λ cloning system has therefore been especially useful for the construction of gene libraries, and organisms from which recombinant λ DNA gene banks have been established include rat (Sargent *et al.*, 1979), yeast (Struhl *et al.*, 1976), *Drosophila melanogaster* (Maniatis *et al.*, 1978), mouse (Blattner *et al.*, 1978) and human (Lawn *et al.*, 1978).

C Cosmids

Cosmids are cloning vehicles derived from plasmids which also contain the λ phage DNA cohesive ends (cos) (Collins and Brüning, 1978; Collins and Hohn, 1979). Only a small region near the cos site is recognized by the λ phage packaging system, both *in vivo* and *in vitro*. A DNA containing two cos sites in the same orientation separated by 35—53 kb can therefore be packaged *in vitro* as described below and can be transduced into *E. coli* at high efficiency (Fig. 4, Feiss *et al.*, 1977). Inside the bacteria the cosmid circularizes (in the same way as does λ DNA), but because the cosmid does not contain all the essential λ phage genes, it is not able to go through a lytic cycle. Instead, cosmids resemble plasmid vectors in usually

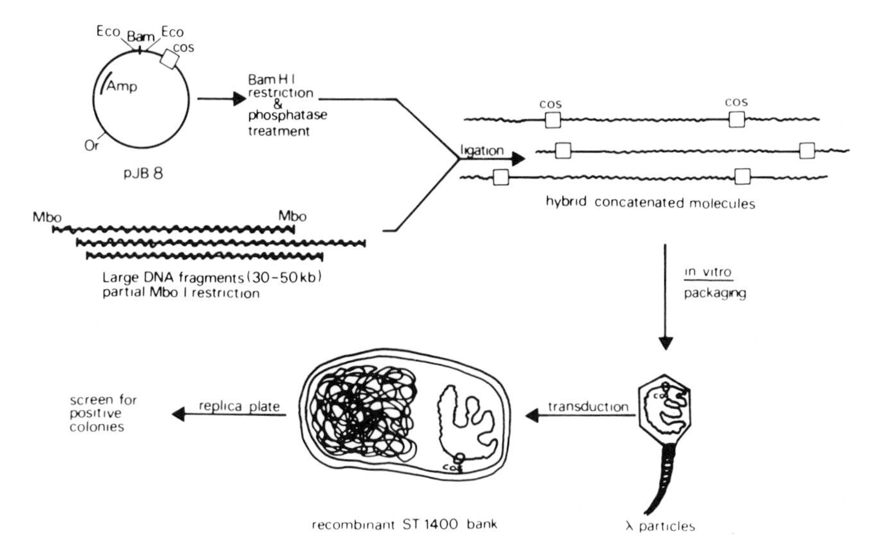

Figure 4 Illustration of a cosmid cloning procedure. The cosmid vector pJB8 is digested with the restriction enzyme Bam HI and treated with phosphatase (see text). Donor DNA is partially degraded with the enzyme Mbo I and fragments 30—50 kb in size isolated by sucrose gradient centrifugation. Digestion with the enzymes Mbo I and Bam HI results in the same overlapping ends on the DNAs, so donor and vector DNA can be joined in a "sticky end" ligation (see text). Hybrid concatamers with cos sites situated approximately 35—50 kb apart, and in the same orientation, can be packaged *in vitro* in λ phage particles and the DNA transduced at high efficiency into an *E. coli* strain containing λ phage receptors (e.g. ST 1400). Bacteria containing a cosmid will be resistant to the antibiotic ampicillin and these can be screened for desired genes using methods described in the text.

conferring resistance to antibiotics; this is used for phenotypic selection. Since cosmids only need to contain a replication origin, selective marker(s) and cos site, they can be small, frequently less than 5 kb in size. The upper size limit of DNA fragments that can be inserted in cosmids and packaged into λ phage particles is therefore approximately 45 kb, much larger than it would be possible to clone in λ or plasmid vectors. The size selection imposed by the *in vitro* packaging system also leads to a minimum size for cloned DNA: if the cosmid to be packaged is 5 kb, the inserted DNA must be at least 30 kb long. The cosmid cloning system therefore offers an efficient system for cloning large DNA fragments, with a low background of non-recombinant colonies, since cosmid dimers are too small to be packaged.

Table 4 summarizes the features of a number of cosmid vectors. The cosmids pJC74, pJC74-58 and pJC720 are derivatives of plasmid Col E1 (Collins and Brüning, 1978). Cosmids pHC79 and pJC81 are

Table 4 **Characterization of some cosmid vectors.**

Cosmid	Useful cloning sites	Selection markers*	Site at which insert inactivates marker	Molecular weight (kb)	Cloning capacity (kb)	Reference
pJC74	Bgl II, Eco RI, Sal I, Bam HI	Amp^R, El imm	—	16	19–36	Collins and Brüning (1978)
pJC75-58	Bam HI, Bgl II, Eco RI	Amp^R	—	12	23–40	Collins and Brüning (1978)
pJC720	Hind III, Xma I	El imm, Rif^R	Colicin production: Xma I	24	11–28	Collins and Brüning (1978)
pHC79	Eco RI, Hind III, Sal I, Bam HI, Pst I, Pvu I, Xma III, Sph I	Amp^R, Tet^R	Amp: Pst I, Pvu I, Tet: Hind III, Sal I, Bam HI, Sph I	6	29–46	Hohn and Collins (1980)
pJB-8	Eco RI, Hind III, Sal I, Bam HI, Sph I	Amp^R	Amp: Pst I, Pvu I	5	30–47	Burke and Ish-Horowich (unpublished)
pOPF1	Bam HI, Kpn I, Cla I Pvu I, Bgl III, Hpa I	Amp^R TK (in mammalian cells)	Amp: Pvu I TK Bgl III	8	27–44	Lund and Grosveld (unpublished)
MUA-3	Pst I, Eco RI, Ava II, Bal I Pvu I, Pvu II, Xma III	Tet^R	—	4.5	31–48	Meyerowitz *et al.* (1980)

*Ampicillin resistance (Amp^R); colicin production (EI imm); rifampicin resistance (Rif^R); tetracycline resistance (Tet^R); thymidine kinase (TK).

derived from plasmid pBR322 (Hohn and Collins, 1980). Cosmid Homer was constructed by inserting a Bgl II fragment from λ DNA containing the cos site into the Bam HI site of plasmid pAT153 (P. Rigby, personal communication). This is relatively easy because Bgl II and Bam HI give similar overlapping ends, which makes the joining of the two fragments straightforward. This, however, leads to loss of the Bam HI sites. A Bam HI site was therefore created by inserting a small, synthetic Eco RI linker containing a Bam site (see below) into the Eco RI site of Homer, giving rise to pJB-8 (J. Burke and D. Ish-Horowich, personal communication).

Recently, cosmid derivatives of pJB-8 containing the herpes simplex thymidine kinase gene have been constructed (F. Grosveld and T. Lund, unpublished). An advantage of these cosmids is that the cloned DNA can be introduced directly into eukaryotic cells that lack thymidine kinase (unpublished observation). Eukaryotic cells containing the recombinant DNA will produce thymidine kinase and can, therefore, be selected for by plating in HAT medium (see Section VII and article by Wickens and Laskey, this series, Vol. 1).

The advantage of using cosmids as cloning vectors, when compared with plasmids or λ phages, is that neither of the latter systems allows efficient cloning of DNA fragments larger than 25 kb, while cosmids can be used for fragments up to approximately 45 kb in size. So far, however, the efficiency of cloning using cosmids has not equalled that of λ vectors. We have observed efficiency with the λ cloning system (λ Ch21A or λ 1059) of 10^5-10^6 clones/μg DNA and with cosmids (pJB8 or pOPF) of 10^4-10^5 clones/μg DNA. A reason for the low efficiency might reflect the difficulties in making large clonable DNA fragments. For example, when DNA fragments of an average size of 30—40 kb are made by partial restriction enzyme digestion (which is the most straightforward method for construction of gene libraries), the starting DNA must be large. Extreme care has to be taken when preparing the starting DNA to prevent shearing or DNase degradation. Furthermore, deletions can occur in cosmid recombinant DNAs, a problem that can be minimized by using a rec A^- *E. coli* host (our unpublished results).

Cosmids have been used to construct gene libraries from *E. coli* (Collins and Brüning, 1978), yeast (Hohn and Hinnen, 1980), *Drosophila* (Hohn and Collins, 1980) and humans (Grosveld *et al.*, manuscript submitted).

D Other vectors

Other cloning vectors are available (e.g. phage M13, virus SV40 or plant Ti plasmids), but none of these have so far proved useful for constructing gene libraries. In phage M13, deletions in large cloned DNA fragments are frequent and probably occur during replication of the single stranded DNA (Herrmann *et al.*, 1980). Fragments that can be cloned in SV40 and propagated as virus stocks have a maximum size of 2—3 kb, due to the small size of this virus and, so far, no suitable plant-cloning system has been developed where a large number of recombinant DNA-containing cells can be obtained and screened.

IV Joining of DNA molecules

Before reintroduction into cells, the vector and donor DNA have to be joined. This is nearly always done *in vitro* using either DNA ligase from *E. coli* or from phage T4-infected *E. coli.* Both enzymes covalently link a deoxyribonucleotide 5′ phosphate with a 3′-OH group (Gumport and Lehman, 1971). Most restriction enzymes create cohesive ends of from 1 to 4 nucleotides. When vector and donor DNA are cleaved by the same enzyme, or enzymes that generate the same sticky ends, the DNA termini can be annealed and covalently linked by DNA ligase. These reactions are usually done at 12—15°C (Dugaiczyk *et al.*, 1975), a temperature that is a compromise between the optimal activity of the ligase (around 37°C) and the stability of annealing of the DNA sticky ends.

It is not always possible to have similar sticky ends on vector and donor DNA. For example, this is the case when:
(a) the restriction enzyme creates blunt ends;
(b) the DNAs are cut with enzymes that create different, non-complementary sticky ends;
(c) the DNA fragments are prepared by mechanical shearing, cDNA synthesis or chemical synthesis.

In cases where the DNAs have blunt ends, T4 DNA ligase can be used to join these under defined conditions (Sgaramella *et al.*, 1970). If the DNAs do not have blunt ends, these can be created by treatment with the single strand specific S1 nuclease (Vogt, 1973). However, the efficiency of blunt end ligation is lower than that of cohesive end ligation (Sugino *et al.*, 1977). A variation of blunt end ligation has been to ligate so-called synthetic DNA linkers to the vector and/or donor DNA (Rothstein *et al.*, 1979). The linker is usually a short DNA fragment which contains the recognition sequence for one or more restriction enzymes, which is not present on the DNA to which the linkers will be ligated. The ends of the linkers and recipient DNA are first made blunt with S1 nuclease. After blunt-end ligation of the linker to the DNA, the ligated DNA is then cut with the corresponding enzyme, which cleaves only the linker, then used in cohesive end ligation. Using the linker technique, the inserted DNA can often be excised from the recombinant molecule by use of the same restriction enzyme.

An alternative method used to link DNA molecules is homopolymer tailing (Jackson *et al.*, 1972). The enzyme terminal nucleotide transferase (Bollum, 1962) is able to catalyse the addition of nucleotides to the 3′ end of a DNA strand. For example, adenine (A) can be added to the vector molecule; thymidine (T) to the donor DNA. When annealed, only A—T duplexes form (preventing self-ligation of either vector or donor DNA) and the annealed DNA can be directly introduced into bacteria where nicks and gaps are filled *in vivo*. The A—T base-paired stretch has a low melting temperature and inserted DNA can often be recovered from recombinants by denaturing the A—T duplex and exposing the single stranded DNA to S1 nuclease degradation (Hofstetter *et al.*, 1976). Similarly, it is also possible to use C—G tailing. As shown in Fig. 5, an advantage with C—G tailing is that conditions can be chosen so that Eco RI or Pst I restriction enzyme sites are reconstructed during the cloning procedure (Rougeon and Mach, 1977; Villa-Komaroff *et al.*, 1978).

It is often of interest to be able to optimize the ligation reaction in order to construct the desired concatamer molecules. This can be done using homopolymer tailing, but can also be done by use of phosphatase from *E. coli* or calf intestine. These enzymes remove the 5′ terminal phosphates from DNA and thereby make it impossible for the DNA ligase to close a nick in the DNA strand (Ullrich *et al.*, 1977). By phosphatase treatment of the DNA molecules (most often the vector is treated) self-ligation or recircularization is prevented. Prevention of these reactions is important in cloning experiments, where reconstruction of the vector leads to formation of a plaque or antibiotic-resistant bacterial colony (depending on the type of vector used), which would significantly increase the background of non-recombinant "clones". A high background complicates subsequent screening, because more plaques/colonies must be examined to isolate a particular clone.

The vector/donor DNA ratio during ligation is of importance in cloning experiments. In the case where λ or a cosmid is used as vector, and the λ packaging system used for transduction, the aim is to form concatamers where both donor DNA ends are joined to vector molecules (Fig. 4). This is favoured by high molar vector/donor DNA ratio. When plasmids are used as cloning vehicles, the recombinant molecules should be a circle consisting of a single vector and a donor DNA fragment. Formation of this type of

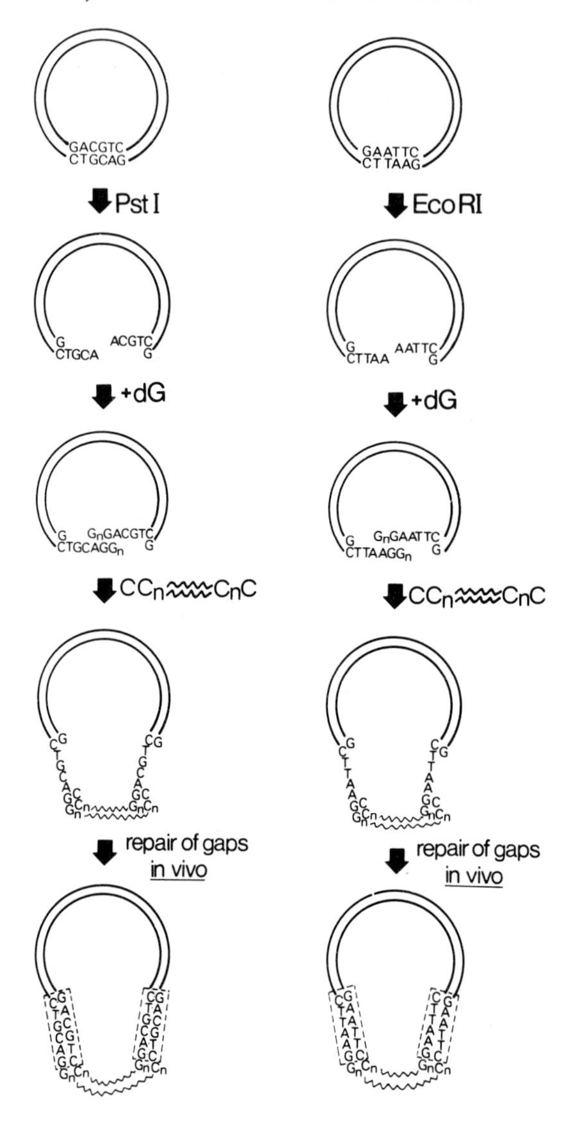

Figure 5 Reconstruction of Pst I and Eco RI restriction enzyme sites using C—G tailing. The vector is digested with either Eco RI or Pst I and dGMP residues are added to the 3′ termini using the enzyme terminal transferase. In a similar reaction dCMP residues are added to the donor DNA fragments. Vector and donor DNA is then hybridized and introduced into bacterium. Gaps are repaired *in vivo* and the Eco RI sites or Pst I sites are restored.

molecule is favoured by a 1:1 molar ratio of vector/donor DNA during ligation. Phosphatase treatment of DNA makes it possible to vary the vector/donor DNA ratio to optimize formation of the wanted chimaeric DNA molecule, without increasing the background significantly. The total DNA concentration in the *in vitro* ligation reaction also influences the type of molecules formed. Low concentrations (less than 20 μg DNA/ml) tend to favour circularization, due to lack of intermolecule interaction, whereas high DNA concentrations (more than 300—400 μg DNA/ml) favour formation of long DNA concatamers.

V Transfer of DNA into bacterial host cells

The actual cloning step is based on the transfer of recombinant DNA to a host cell, followed by the isolation of one species of recombinant initially derived from a single phage or transformed cell. The host, therefore, must be able to replicate the foreign DNA, and express a phenotype conferred to it by the recombinant molecule that can be selected for or recognized by a screening method. This description of DNA transfer into host cells is limited to *E. coli*, since in almost all recombinant DNA research, mutants of *E. coli* K12 have been used. These hosts lack a restriction system and do not degrade (restrict) foreign unmodified DNA when it is introduced in the cell. There are, however, other bacterial systems, such as *B. subtilis*, available (Keggins *et al*., 1978; Ehrlich, 1977; Bernhard *et al*., 1978; Gryczan *et al*., 1978; Löfdahl *et al*., 1978).

For the transfer of recombinant DNA into *E. coli* there are basically two methods in use: transformation—transfection and transduction.

A Transformation—transfection

The procedures for transformation (transfer of plasmids) and transfection (transfer of phage) most commonly used are minor modifications of that of Mandel and Higa (1970) based on a $CaCl_2$ treatment that renders *E. coli* susceptible to DNA uptake. The recombinant DNA is simply mixed with $CaCl_2$-treated bacteria and the mixture incubated on ice, followed by a short incubation at 42° C (heat shock). In a transfection the bacteria are then immediately plated in agar. Where a phage DNA molecule has been taken up by a bacterium, it will be expressed and give rise to infectious phage particles. Upon lysis of the host cells, the newly synthesized particles will infect neighbouring bacteria in the agar and induce another round of phage synthesis and lysis, eventually leading to a clear plaque of lysed bacteria in the bacterial lawn. Scoring of the number of plaques is a measure of the number of transfected λ DNA molecules. In a transformation experiment the bacteria are first incubated in a non-selective broth to allow time for the expression of the newly acquired phenotype (e.g. ampicillin or tetracycline resistance). The bacteria are then plated on selective medium (ampicillin or tetracycline). Only the bacteria that have taken up a plasmid with a drug resistance marker will be able to grow and form a colony of bacteria. Consequently, the number of colonies is a direct measure of the number of absorbed plasmid molecules. For some host strains special methods have been developed either to obtain a high transformation efficiency (Kushner, 1978)(*E. coli* SK1592) or because the strains are very sensitive to lysis (Inoue and Curtiss, 1977)(*E. coli* X1776).

In a normal experiment about 10^4—10^6 plaques for λ DNA and 10^6 colonies for plasmid pBR322 can be obtained per microgram of input DNA. For *E. coli* X1776, frequencies as high as 10^8 can be obtained (D. Hanahan, personal communication). In recombinant DNA transfers these numbers will usually be 10^2—10^4-fold lower for two reasons. Firstly, ligation of vector and insert DNA is generally inefficient and will lead to a lower number of functional plasmid or phage molecules. Secondly, vector molecules containing an inserted DNA fragment are larger and therefore have a lower transformation—transfection efficiency than the vector DNA. From comparison of the numbers, it is immediately apparent that precautions should be taken to suppress a background of transformants containing phage or plasmid DNA without an inserted fragment. This suppression can be achieved by preventing the formation of circular vector molecules without an insert (as discussed earlier in this chapter in "joining of DNA molecules"), or by killing bacteria containing only the original vector after the transformation by cycloserine enrichment (Rodriguez *et al.*, 1976). This method is based on the inactivation of a plasmid gene by the insertion of a foreign fragment. For instance, when the foreign DNA is inserted in the tetracycline resistance gene, that gene is inactivated, while the second resistance marker (ampicillin) remains intact. A transformed culture is then grown up in ampicillin-containing medium, which selects all cells carrying a plasmid. The enriched culture is then diluted and grown in tetracycline during which all cells containing an insert in the pBR322 tetracycline resistance (Rodriguez *et al.*, 1976) gene will cease growing. Cycloserine is then added which will rapidly lyse all growing cells and the remaining cells can be rescued by centrifugation. After this treatment, between 90% and 100% of these rescued cells will contain plasmids that have acquired insertions (Rodriguez *et al.*, 1976).

B Transduction

Transduction procedures are based on an *in vitro* system to package either λ or cosmid recombinant DNA into infectious λ phage particles and provides a method to transfer large recombinant DNA molecules with a high efficiency into *E. coli*. From the original observations of Becker and Gold (1975) and Hohn (1975) on the packaging of λ DNA into particles during λ morphogenesis, procedures were developed so that only exogenous DNA could be efficiently packaged into particles (Sternberg *et al.*, 1977; Hohn and Murray, 1978). It involves the preparation of complementary extracts from two *E. coli* lysogens, each of which is blocked in one step of λ particle morphogenesis: either nonsense mutations in λ capsid genes A and E (strains NS 1128 and 433, N. Sternberg), or mutations in λ genes D and E (BHB 2690 and 2688, B. Hohn). An extract from the mutant E strain lacks the major capsid protein E but contains all other head proteins in a soluble form. This extract can therefore complement the second extract which contains empty precursor particles to complete the λ morphogenesis. Additional mutations in these prophages and in the host strains provide efficient induction of prophage and prevention of lysis of the induced bacteria, packaging of endogenous induced prophage λ DNA and recombination in the packaging extracts. The most efficient procedure developed to date (Pirotta, unpublished results), which is a combination of the procedures referred to above, will give as many as 5×10^8 plaques per microgram of packaged wtλ DNA. This number will drop 10^2—10^4-fold for recombinant phage or cosmid DNA for the same reasons as described above, plus the fact that only molecules ranging in size from 75% to 105% of the size of λ DNA can be packaged efficiently (Feiss *et al.*, 1977). It is worth noting that there is an additional effect of the size range limitation when either single recombinant λ phage, or complete libraries, are grown in large cultures. During growth the recombinant λ molecules are repackaged every generation, and for λ recombinants that have a size at the borders of the packaging range, this will result in both a low yield (Feiss *et al.*, 1977) and selection for certain recombinants in the case of whole libraries. Since cosmids are only packaged once (at transduction) there is no such restraint.

VI Screening clones

The final step in a cloning procedure is the screening of a population of recombinant bacterial clones, or phage plaques, for the ones containing the foreign DNA sequence of interest. The screening procedures fall in two classes: direct selection of plasmids carrying fragments that give a selectable phenotype to the host, and indirect selection by detecting which recombinant *E. coli* contains nucleotide sequences or gene products of the fragments of interest.

A Direct selection by phenotype

This screening method is based on the suppression or complementation of an *E. coli* mutation *in vivo* by transformation with foreign DNA. Many *E. coli* strains with well characterized mutations are available and in many of these suppression or complementation can be obtained by low levels of expression of the gene product coded for by the foreign DNA. The requirements for this are that the cloned fragments are large enough to contain a complete functional unit and that the cloned gene can be expressed in *E. coli.* Using a variety of mutants and procedures a number of genes from *E. coli* (Clarke and Carbon, 1975, 1976), yeasts (Struhl *et al.*, 1976; Clarke and Carbon, 1978) and fungi (Ratzkin and Carbon, 1977; Vapnek *et al.*, 1977) have been isolated in this manner.

It is much more difficult to apply these methods to the isolation of eukaryotic genes for two reasons: many gene products from eukaryotes will not complement or suppress *E. coli* mutations and many genes from eukaryotes contain intervening sequences, which will prevent the expression of a functional gene product since *E. coli* does not contain the necessary splicing mechanism. In some cases these problems can be bypassed by using reverse transcripts of mRNAs, to give recombinants that *E. coli* will transcribe and translate into protein. Such recombinants are usually identified by immunological or biological assays, as described below, though in some instances direct selection is possible. For instance, the mouse dihydrofolate reductase gene was cloned in *E. coli* by phenotype selection, because the gene conferred trimethoprim resistance (Chang *et al.*, 1978).

B Indirect selection

A range of screening procedures has been developed to detect the recombinant of interest by probing either for the new DNA sequences conferred to it, or a particular protein product coded by the inserted fragment, or a combination of both.

C Nucleic acid hybridization

The most widely used method to identify clones within a library carrying specific fragments is by DNA—DNA or DNA—RNA hybridization with radioactive gene-specific DNA or RNA probes. The method to screen many recombinant colonies *in situ* with a radioactive probe which was originally developed by Grunstein and Hogness (1975) and since modified for high colony density by

Hanahan and Meselson (1980) is illustrated in Fig. 6. The transformed bacteria are plated on nitrocellulose filters on agar dishes and colonies grown. These are then replica plated onto new filters which are grown while the master plate is retained. The replica filters are then removed from their plates and the DNA in each colony fixed to the filter by alkaline denaturation and drying of the filters at high temperature. The probe radioactivity, labelled RNA or DNA, is then hybridized to these filters and the result monitored by autoradiography. Whenever a colony contains sequences complementary to the probe, a spot will appear on the film and, by determining its position, the positive colony can be picked from the master filter or plate. A modification of this colony screening procedure was developed by Benton and Davis (1977) for the analysis of recombinant phage plaques. In this procedure phage particles with free unpackaged λ DNA that is always present in the plaques are transferred to a nitrocellulose filter by putting a filter on top of the bacterial lawn.

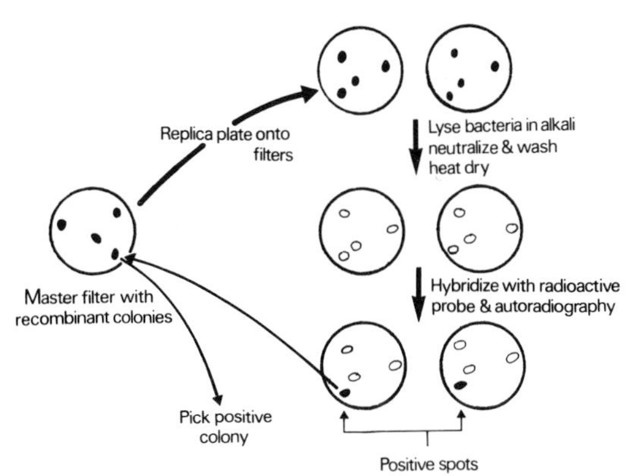

Figure 6 Diagram of a procedure to screen recombinant clones in *E. coli* with radioactive nucleic acid probes.

The phage DNA bound to the filter is then denatured and neutralized to bind it *in situ* and fixed to the filter by drying at high temperature as in the colony hybridization procedure. The phage procedure generally has a better signal to background ratio than the procedures used to screen plasmid-containing bacterial colonies. This is probably because more recombinant DNA and less cellular debris is fixed on the filter. A combination of both procedures can be used with *E. coli* that carries thermo-inducible λ lysogen (Cami and Kourilsky, 1978). The initial part of this procedure (replica plating) is carried out as in a colony hybridization. The replicas are then incubated at the non-permissive temperature which induces the λ lysogen;

thereafter packaging of the recombinant DNA and cell lysis occurs. The fixing of the DNA to a filter can then be carried out as for plaque hybridization.

Providing that a probe is available, the advantage of the hybridization method is its independence of expression of the foreign DNA by the *E. coli* host. Both *in vivo* and more usually *in vitro* labelled RNA and DNA probes can be used (Grunstein and Hogness, 1975; Maniatis *et al.*, 1978; Villareal and Berg, 1977). When a probe is not available as a purified fragment or a very abundant species (like ribosomal RNA) in a total population, a differential hybridization (St John and Davis, 1979; Taniguchi *et al.*, 1979; Hoeijmakers *et al.*, 1980) can sometimes be applied. This method is based on the fact that certain genes can be induced specifically, or are present in only one of two homologous species, or are only expressed during certain stages of development. In all of those cases, two populations of mRNA can be obtained: one with the mRNA of interest as part of the total mRNA population and one without. The recombinant population is then screened in parallel with the two preparations of total mRNA (or its cDNA) as probe. When the mRNA of interest is present, all colonies containing recombinant DNA will be positive on the film. When the mRNA of interest is absent all colonies will be positive with the exception of those containing the gene of interest. These can then be picked from a master plate for further analysis.

Chemical synthesis of DNA is another possible way to provide radioactive probes for screening when natural homologous DNA or RNA fragments are not available. This approach can be applied when (parts of) the amino acid sequence of the gene product of interest is known. From this information a short protein sequence is selected which contains amino acids with unique codons (Met and Trp) or the least redundancy in codons. An oligonucleotide corresponding to the sequence of these codons is then synthesized, labelled and used as a probe (Montgomery *et al.*, 1978; Noyes *et al.*, 1979).

D Hybrid-arrested and hybrid-selected translation

Screening by hybrid-arrested translation can be applied in the case of a gene that codes for an abundant mRNA (Paterson *et al.*, 1977; Hastie and Held, 1978). It is based on the fact that an mRNA will not direct the synthesis of a polypeptide in a cell-free translation system when it is hybridized to DNA. The DNA is prepared from pools of transformed *E. coli* or phage plaques and hybridized to a total mRNA population under conditions where DNA—RNA hybrids are more stable than DNA—DNA hybrids. The total mixture is then incubated in a cell-free translation system and the results compared with the translation of non-hybridized mRNA, and a denatured control. Inhibition of translation of the protein of interest will then identify the pooled *E. coli* or plaques that contain the DNA of the gene of interest. This pool can then be subdivided into smaller pools, or a matrix of pools and the experiment repeated until specific clones can be identified.

Hybrid selected translation is a much more sensitive method and also applicable to low abundance mRNAs (0.1% of total mRNA). It is based on the same principle, but in this method the presence, rather than the absence, of a particular active mRNA in a translation assay is obtained. The DNA from pools of recombinants is hybridized with an mRNA population, either with the DNA fixed to a solid support (Sobel *et al.*, 1978; Nagata *et al.*, 1980) or in solution (Woolford and Rosbash, 1979). The hybridized mRNA is isolated by elution from the bound DNA, or by isolation of the hybrid from the total RNA by column chromatography. The hybridized RNA is then recovered and translated *in vitro*. The protein product is then usually identified by gel electrophoresis, but in some cases by biological activity (Nagata *et al.*, 1980). Once a positive recombinant pool is established, this can be subdivided until one or several single recombinants are scored positive.

E Immunochemical methods

Immunochemical detection of recombinant clones can be applied successfully when the foreign DNA is at least partially expressed by *E. coli*. An advantage of these methods over direct selection methods (see above) is that the foreign gene product does not have to confer any selectable phenotype to the host. On the other hand, it does require the availability of specific antibodies. Some relatively insensitive methods have been developed for detection of protein directly in λ plaques or bacterial colonies on agar (Sanzey *et al.*, 1976; Shalka and Shapiro, 1976), but the more sensitive methods involve solid matrix-bound and radioactively-labelled antibodies (Carbon *et al.*, 1978; Ehrlich *et al.*, 1978; Broome and Gilbert, 1978). Moreover, these methods depend on the fact that the immune serum used contains several antibodies that recognize different determinants on the antigen molecule. The antibody or the F(ab) part of the antibody is bound to a solid matrix (Fig. 7). The matrix is brought in contact with the lysed bacterial colonies or pools of colonies and the antigen allowed to bind to the antibody on the matrix. The matrix is then washed and incubated either with *in vitro* labelled antibody, or with unlabelled antibody, followed by *in vitro* labelled *Staphylococcus* protein A that specifically binds to the Fc

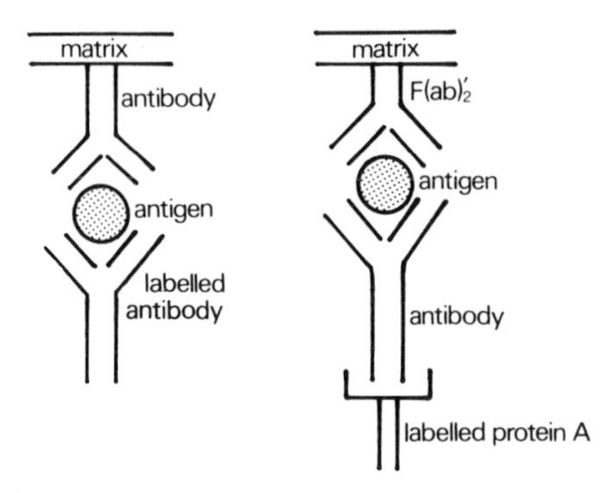

Figure 7 **Diagram of screening with labelled antibodies or protein A.**

portion of the antibody. Positive colonies are detected by autoradiography and picked from a duplicate plate of colonies. Colonies containing hybrids of two proteins (A—B) can also be detected in a two antibody assay. An antibody against part (A) of the hybrid is fixed to the matrix, while an antibody against part B of the hybrid is used in the detection. Since the first antibody only binds A, while the second antibody detects B, only the hybrid A—B protein-containing colony would show up as positive.

VII Cloning in eukaryotic cells

Several cloning procedures in eukaryotic cells have been developed for particular purposes, but many laboratories are at present developing new procedures that could have a more general application. The availability of a general cloning procedure is important, since many genes cannot be identified in *E. coli* unless sequence-specific nucleic acid probes are available.

The introduction of new DNA into eukaryotic cells can be achieved in a variety of ways, but the most widely used procedures are based on the uptake of DNA as part of a calcium phosphate precipitate (Graham and Van Der Eb, 1973), after which the DNA is transported to the nucleus and apparently integrated in the host chromosome to give stable transformants. The DNA of simian virus 40 (SV40) has been the most widely used vector to introduce new DNA into eukaryotic cells, but mostly to introduce identified DNA fragments (already cloned in *E. coli*) into cells to study the expression of particular genes (Upcroft *et al.*, 1978; Mulligan *et al.*, 1979). The disadvantage of SV40 as a general cloning vehicle is the limited capacity of the vector (4.3 kb of added DNA) when used in a lytic cycle of infection in permissive cells, and the absence of a convenient selection procedure when used in non-permissive cells. When it was shown that yeast could be transformed with cloned yeast DNA sequences (Hinnen *et al.*, 1978), it appeared likely that the 2 μm circular DNA of yeast could be used as a cloning vehicle for eukaryotic genes (Sinsheimer, 1977). However, it was subsequently shown that yeast is unsuitable as a host for the general cloning of eukaryotic genes, since it does not express the rabbit β-globin gene faithfully, due to incorrect transcription and DNA splicing (Beggs *et al.*, 1980). This does not apply to the cloning of yeast genes in yeast; it has been shown that yeast can be used as a host to clone and express genes from different yeast species (Dickson, 1980).

The only general procedure that has been used successfully in eukaryotic cells has been a selection procedure involving thymidine kinase deficient (TK^-) mouse cells. These can be transformed to a TK^+ phenotype with DNA fragments from herpes simplex virus (HSV) carrying the HSV TK gene (Wigler *et al.*, 1977, Bacchetti and Graham, 1977; Pellicer *et al.*, 1978). The transformed TK^+ cells can be selected over the TK^- cells using HAT medium, which contains aminopterin (A)(which blocks nucleotide biosynthesis) and hypoxanthine (H) and thymidine (T)(two precursors that enter in the biosynthetic pathway after the blockage but which require thymidine kinase to complete synthesis)(Fig. 8).

The HSV TK gene cloned in a bacterial vector can, therefore, serve as a eukaryotic vector with a selectable marker to introduce and express new DNA in TK^- cells. This was shown for the rabbit β-globin gene which was ligated to the HSV TK gene/plasmid vector, and this recombinant introduced into TK^- mouse L cells. The cells which were selected in HAT were shown to contain and express both the HSV TK and rabbit β-globin genes (Mantei *et al.*, 1979). In a theoretical attempt to clone a gene from total chromosomal DNA, fragments would be ligated to the TK vector and used (directly or after an *E. coli* cloning step) to transform TK^- cells (Fig. 9). TK^+ transformation would be selected in HAT medium and screened for the cells containing the gene of interest, e.g. by immunological techniques. The DNA from positive cells is then extracted and the gene of interest isolated by recloning in *E. coli* (plasmid rescue, see below). To date, there has been no report of the successful cloning of a eukaryotic gene using this type of protocol, although many laboratories are working on such methods with newly developed plasmid-TK and cosmid-TK vectors.

Rescue has been used successfully to isolate the chicken thymidine kinase gene (Perucho *et al.*, 1980), but this involved only direct selection for a TK^+ phenotype and not an additional screening for a non-selectable phenotype. Fragments of chicken chromosomal

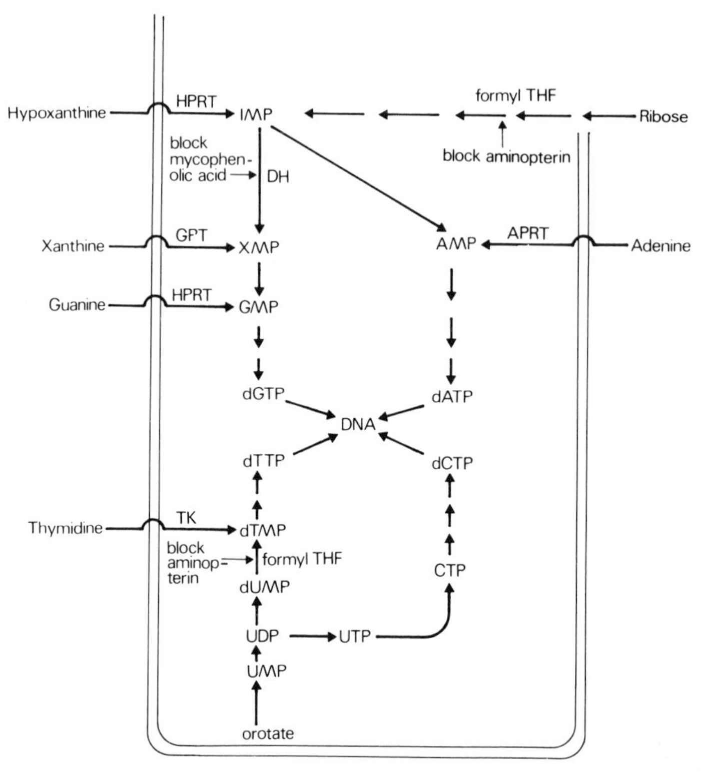

Figure 8 Schematic representation of the nucleotide bisoynthesis pathway. Enzyme abbreviations: APRT, adenine phosphoribosyl transferase; HPRT, hypoxanthine guanine phosphoribosyl transferase; GPT, guanine phosphorybosyl transferase; TK, thymidine kinase.

DNA (which had previously been shown to give stable transformation of cells to TK^+ (Wigler *et al.*, 1978)), were linked to pBR322 DNA and the hybrid molecules introduced in TK^- mouse cells. DNA from a TK^+ transformant was then isolated, cleaved with a restriction enzyme that leaves both pBR322 and the chicken TK gene intact, circularized and used to transform *E. coli* (Fig. 9). Since pBR322 DNA contains a bacterial origin of replication and carries a bacterial drug resistance marker, only those DNA circles containing pBR322 can transform *E. coli* and confer resistance to an antibiotic. Plasmid DNA containing pBR322 and the linked chicken TK gene was then isolated from transformed *E. coli* colonies.

The TK vector system depends on the availability of a suitable TK^- host cell. Since a given gene might only be expressed in certain cells, it may well be necessary to first isolate stable TK^- mutants of a particular cell type. To circumvent this problem, a number of new vector systems are currently under development in order to provide

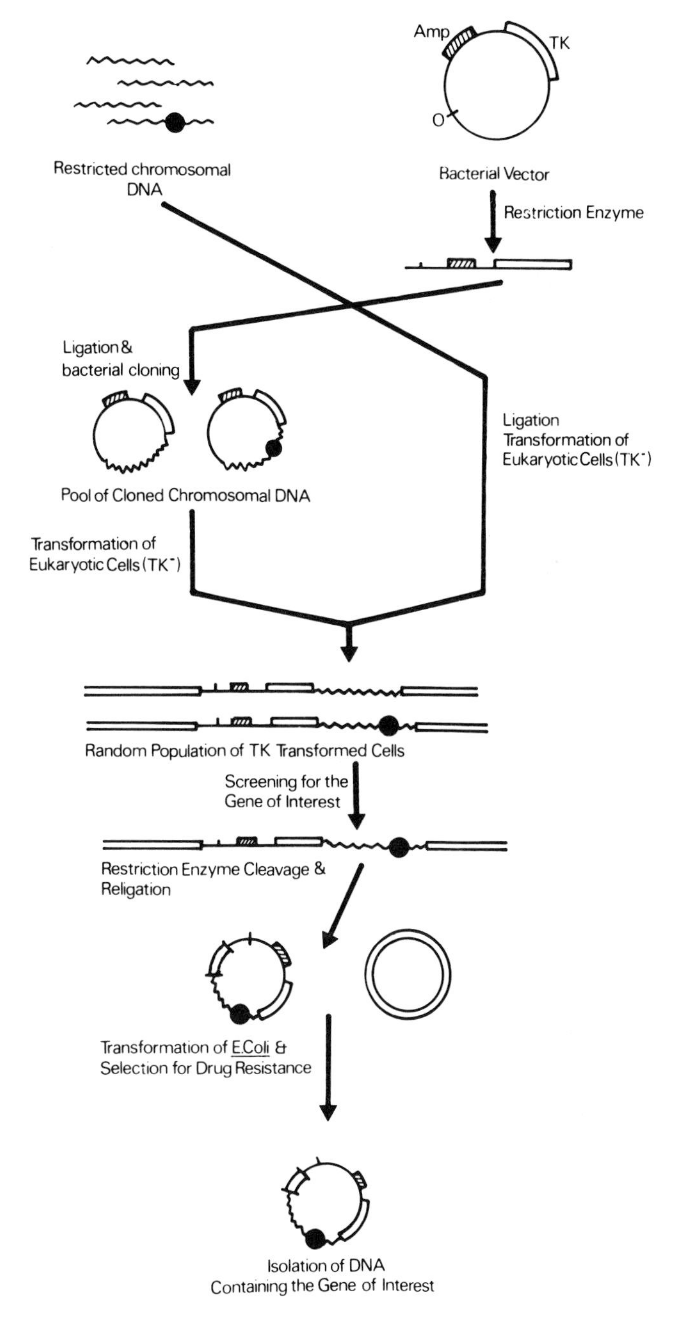

Figure 9 Theoretical scheme for the cloning of a gene using gene rescue in eukaryotic cells. Symbols: hatched box, ampicillin resistance marker; open box, thymidine kinase gene; open ended boxes, host eukaryotic cell DNA, filled circle, the gene to be cloned.

a direct selection system that could be applied to almost any cell type available. One such system (Jimenez and Davis, 1980) involves vectors that can express a kanamycin resistance marker in both bacteria and eukaryotic cells, when the kanamycin derivative drug 2-deoxystreptamin is used in the selective medium. Only bacteria or eukaryotic cells containing a recombinant carrying the resistance marker are then able to grow, since both bacteria and eukaryotic cells are sensitive to this drug. Since essentially all eukaryotic cells are sensitive to this drug, the selection is of almost universal application.

Another system recently described uses a selection based upon blocking parts of the nucleotide biosynthesis pathway in a similar way to that employed for TK (Fig. 8). In this case, a vector is used that can express the *E. coli* enzyme guanosyl phosphotransferase (GPT), an enzyme that is absent in eukaryotic cells (Mulligan and Berg, 1980). Transformed cells are selected in medium containing both mycophenolic acid (MPA), which blocks the conversion of IMP to XMP, and xanthine, which can enter the biosynthetic pathway after the block at GPT. Both these systems have the advantage over the TK system in that they are independent of the availability of mutant host cells and can, in principle, be used in many different cell types.

VIII Stability of library clones

Numerous experiments have shown that most DNA regions examined so far can be isolated without rearrangement and maintained as stable recombinant DNA clones in bacteria. In a few cases, however, specific deletions in cloned DNA have been detected, and shown to be artefacts of the cloning procedure. First, a 4.35 kb deletion was found in a clone containing a rabbit β-like globin isolated from a phage λ library (Lacy *et al.*, 1979). Second, globin clones isolated from a similar human gene library showed deletions in a region between and including part of the $^{G}\gamma$ and $^{A}\gamma$ globin genes (Fritsch *et al.*, 1980). In addition, specific deletions occur at high frequency in two phage λ clones containing human α-like globin genes (Lauer *et al.*, 1980). It is striking that all these deletions occur between regions with a high degree of homology. It is therefore tempting to suggest that these deletions arise by unequal crossing over (either inter or intramolecular) of homologous segments during amplification of the library or further growth of the phage.

We have also found deletions in one cosmid clone containing the human β-like globin genes (unpublished results), although unlike the phage λ clones, cosmid clones which contain the $^{G}\gamma$- and $^{A}\gamma$-globin genes are perfectly stable. Deletions occur at very low frequency when the cosmids are contained in a rec A^- host. We have, however, inserted additional DNA sequences in these cosmid clones in several experiments, which bring the size of the cosmid close to the upper limit for packaging *in vitro* in phage λ particles. After packaging and rescreening, clones containing deletions were found at high frequency. It is not established if this is due to the higher packaging efficiency of a small population of deleted molecules and/or if larger molecules are more likely to undergo deletion (F. Grosveld and H. Dahl, unpublished). These observations emphasize the importance of comparing the cloned DNA fragments with those of genomic DNA, to establish that no rearrangements have occurred during cloning.

IX Applications

A Gene structure

The fact that single genes can be isolated in a pure form by recombinant DNA technology has made possible detailed analysis of gene structure. Since large cloned segments are difficult to analyse in detail (as from a phage or cosmid library), it is frequently useful to localize the gene within the cloned segment prior to further studies. A combination of electron microscopic techniques and the use of restriction endonucleases permits a rapid localization of the structural gene (Table 5). Usually the starting point is a phage or plasmid clone that has been shown to hybridize with a nucleic acid probe — a purified mRNA or a cloned cDNA copy of this mRNA. A useful first step in the analysis of the genomic clone is the construction of a simple restriction map, since this makes it possible to focus on the structural gene. To localize the gene within the genomic clone, the Southern blotting technique is used. In this, the DNA is cleaved with one or more restriction endonucleases, electrophoresed in an agarose gel to separate the DNA fragments on the basis of size, denatured by alkali, neutralized and then transferred to a nitrocellulose sheet. The

Table 5 An outline procedure for the localization and characterization of a gene on a cloned DNA segment

1	Determine a simple restriction map for the large cloned DNA segment to segment the clone into pieces of a few kb.
2	Hybridize Southern blots of a suitable digest with the relevant nucleic acid probe. Determine which fragments hybridize.
3	Perform R-loop analysis in the electron microscope using mRNA and the original cloned DNA or a smaller segment of this.

DNA binds to the nitrocellulose *in situ* to give a filter-paper replica of the DNA separated in the gel. The filter bound DNA is then hybridized to mRNA, or cDNA, to detect the gene sequences: the labelled bands are detected by autoradiography.

A more precise localization is possible using electron microscopy. Usually the R-looping method is used (White and Hogness, 1977). The physical basis of the method probably relies on the observation that DNA—RNA hybrids (or RNA duplexes) consist of an A-type helix, whereas DNA duplexes are in the B form. The altered structure of the helix confers a difference in stability between the two forms such that in aqueous solvents DNA duplexes are generally more stable than the DNA—RNA hybrids. Birnstiel *et al.* (1972) showed that, in solvents such as formamide, DNA—RNA hybrids are more stable than the homologous DNA—DNA duplex; the temperature at which denaturation occurs (Tm) for a DNA—RNA hybrid is about 6°C higher than that of a DNA duplex under these conditions (Casey and Davidson, 1977). To form R loops, therefore, a duplex DNA molecule is annealed with a homologous RNA in a formamide-containing solvent at temperatures close to the Tm of the DNA. The DNA duplex will either partially melt (particularly in A—T rich regions), or breathe, so that transient single stranded regions will be exposed. Since the RNA—DNA hybrid is stable under these conditions, this structure will form, thereby displacing the homologous part of the non-coding DNA strand. These structures can easily be visualized in the electron microscope. If the structural gene is an internal segment of the clone under analysis, then the non-coding DNA strand will be shown as a displacement loop called an R loop (Fig. 10; note that the loop is DNA and not RNA). By using cloned DNA cleaved by a series of restriction enzymes in the R-looping experiments, it is obviously possible to position the structural gene quite accurately on the DNA molecule. In practice, the resolution will be of the order of 100 bp.

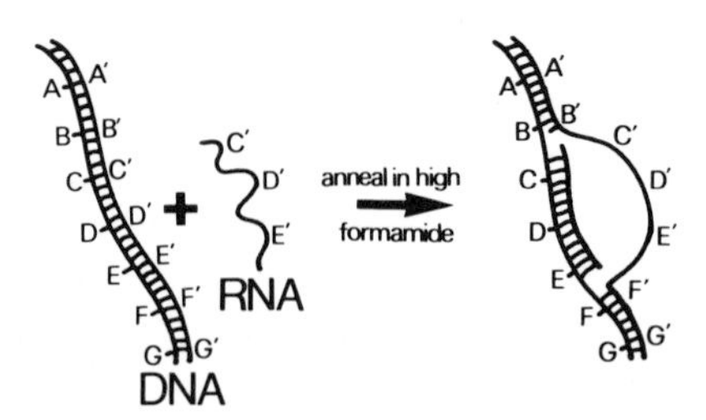

Figure 10 Schematic illustration of the formation of an R loop (see text for details).

It is, of course, also important to determine the orientation of the gene within the DNA segment cloned. To do this, the cloned DNA is cleaved with a restriction endonuclease that cuts within the gene. The DNA is then blotted into nitrocellulose and hybridized with segments of the cDNA probe that contain either the 5′ or the 3′ regions of the gene (these are, of course, also prepared by cleavage of the cDNA with a restriction endonuclease). The 5′ and 3′ probes will detect the corresponding regions of the genomic DNA. Since a restriction map is already available, this enables the orientation of the gene to be elucidated.

The methods described above make it possible to determine the structural organization of the gene. The fact that many eukaryotic genes are split, consisting of non-contiguous blocks of mRNA-coding sequences (or exons) separated by intervening sequences (or introns), was shown by R looping (Tilghman *et al.*, 1978b) and Southern blotting with genomic DNA (Jeffreys and Flavell, 1977; Breathnach *et al.*, 1977; Doel *et al.*, 1977). In this type of gene, the mRNA will hybridize to non-contiguous segments of the cloned DNA. This can be shown by hybridizing mRNA or cDNA probes to Southern blots of the cloned DNA or to total chromosomal DNA. In either case, it is of course essential that the restriction map of the cDNA is known. Although using chromosomal DNA is technically somewhat more demanding (the sensitivity of the hybridization needs to be about 10^4—10^5-fold higher than that required with cloned DNA), it does have the advantage that the possibility that the cloned DNA has undergone rearrangements does not need to be considered. This was very useful when the first intervening sequences of cellular DNA were discovered in the globin (Jeffreys and Flavell, 1977; Tilghman *et al.*, 1978b) and ovalbumin genes (Breathnach *et al.*, 1977; Doel *et al.*, 1977).

R-looping is a very rapid way of determining whether a gene contains large introns. If an intron is present, the R loop will be multiple instead of single (Fig. 11); the number of loops is a minimum estimate of the number of mRNA coding blocks in the gene. In practice, there may well be many more coding blocks because short intervening sequences may be missed, or even prevented from re-forming a DNA duplex by the DNA—RNA hybrids.

When the gross structure of the gene has been established, the fine structure remains to be determined. Of particular importance here is the position of the 5′ and 3′ termini of the gene and the precise boundaries and DNA sequence of the introns. This is performed by determining the nucleotide sequence using either the chemical methods of Maxam and Gilbert (1980), or the enzymatic methods of Sanger *et al.* (1977). Comparison of the gene sequence with that

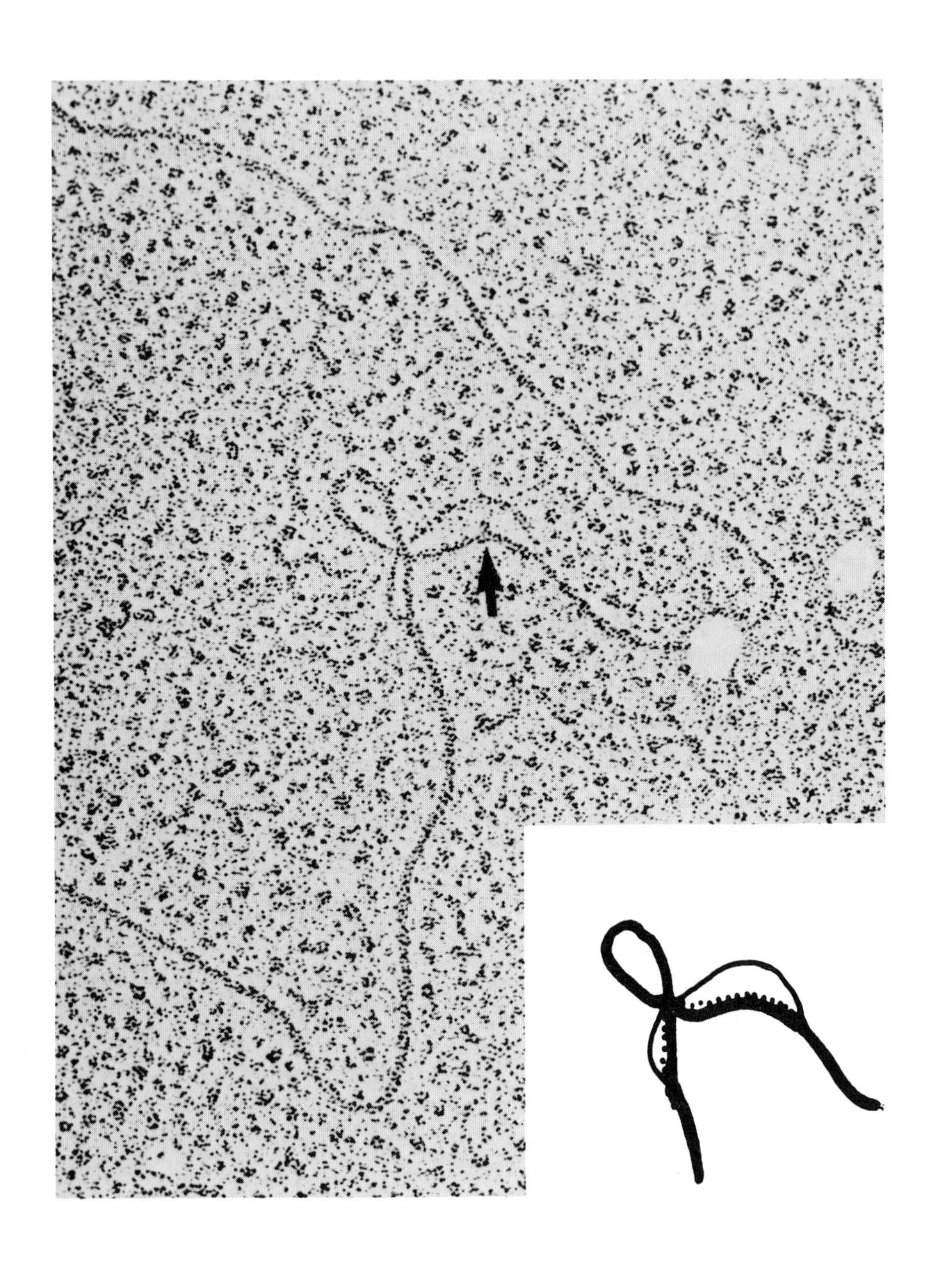

Figure 11 R loop formed between mouse β-globin mRNA and the "split" mouse β-major globin gene. The arrow points to a small loop that is probably the small intron of the β-globin gene. The insert shows the interpreted structure of the R loop. The dotted line indicates the RNA and the thin line the displaced DNA strand. From Tilghman *et al.* (1978b).

of the mRNA and protein sequence is almost essential. This is commonly done by comparing the DNA sequence of the cloned cDNA with its genomic counterpart. In practice, most cDNAs lack the 5′-terminal sequences of the mRNA (see Williams, this series, Vol. 1) so that the 5′ sequence of the mRNA has usually to be determined by direct RNA sequencing.

B Introns

From the approaches outlined above, we have learned a great deal about the structure of eukaryotic genes. First and foremost, many eukaryotic genes are split; comparison has pointed to several general conclusions:

1 *The position of introns in homologous genes is constant*

This has been best studied in the haemoglobin system where it was shown that the positions of the two introns in rabbit (Jeffreys and Flavell, 1977; Van Den Berg *et al.*, 1978), mouse (Tilghman *et al.*, 1978b; Konkel *et al.*, 1978; Van Dan Berg *et al.*, 1978) and human (Flavell *et al.*, 1978; Lawn *et al.*, 1978) β-globin and those of the mouse (Leder *et al.*, 1978) and human (Lauer *et al.*, 1980) α-globin genes are constant. The α- and β-globin genes are believed to have been derived from a common ancestral gene by a duplication event which must have occurred about 500 million years ago (Jeffreys, this volume). It is most likely, therefore, that the two intervening sequences were also present in the same position in the original globin gene and that split genes are a very old phenomenon. The alternative explanation, namely that both introns were added to these genes more recently is less likely, but remains to be formally excluded.

In at least one interesting case, however, relatively recent losses of an intron seem to have occurred. The insulin genes of chicken (Perler *et al.*, 1980) and man (Bell *et al.*, 1980) have two intervening sequences. The rat has two insulin genes; one of these has a similar structure with two introns, but the other gene has lost the second intron (Lomedico *et al.*, 1979).

2 *Intron DNA sequences are usually less conserved than mRNA coding sequences*

Sequence comparison of several related globin genes from different species shows that, in general, intron sequences diverge more rapidly than the coding sequences (e.g. Van Den Berg *et al.*, 1978). An exception to this is the intron—exon junctions where significant sequence conservation is evident. Breathnach *et al.* (1978) and Catterall *et al.* (1978) have compared the sequences at these junctions and shown that the 5′ (donor) splice junctions have the sequence GT and the 3′ (acceptor) junctions have the sequence AG (Fig. 12). These conserved sequences are presumed to be involved in the recognition of the splice sites of the pre-mRNA by the splicing enzymes.

The study of processed RNAs (in the case of β-globin mRNA present in early erythroid cells) shows that splicing is an extremely precise process. Since the dinucleotides GT . . . AG occur frequently in eukaryotic mRNAs, it is clear that the recognition sequences for splicing enzymes must be longer than these dinucleotides. Comparison of a large number of splice sites suggests preferred donor and acceptor splice junctions (Fig. 12).

Recent evidence suggests that a small nuclear RNA (snRNA) may well be involved in the splicing reaction (Lerner *et al.*, 1980). U1 RNA is the most abundant snRNA and sequence comparison shows that the 5′ end of U1 RNA is complementary to the intron—exon junction sequences (Fig. 13). U1 RNA is present in nucleoprotein complexes (snRNP).

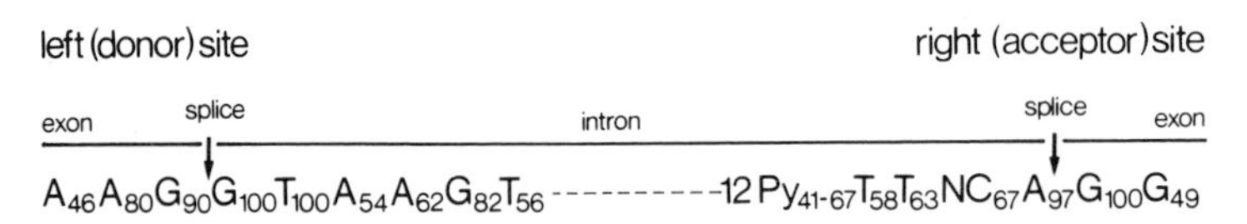

Figure 12 Consensus sequence for the donor and acceptor splice junctions. The sequences have been compared and the number shown at each residue indicates the frequency (as a percentage) of the given nucleotide at that position. From Lewin (1980) with permission of MIT press.

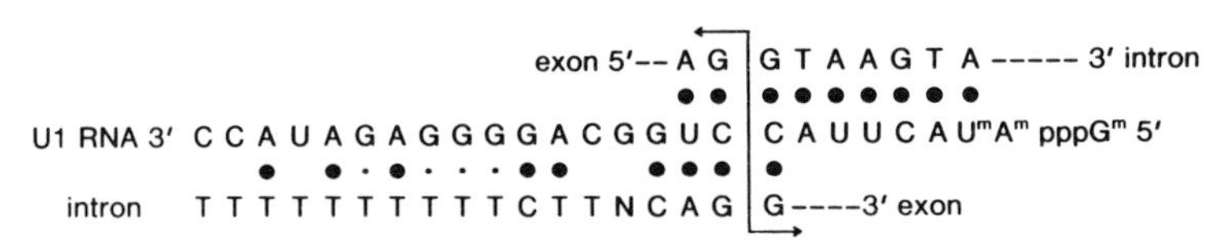

Figure 13 Sequence homology between U1 RNA and splice junctions. From Lewin (1980) with permission of MIT press.

3 *Introns commonly separate functional domains of proteins*

In the discussion of how split genes evolved an early and attractive idea expressed is that the exons encode functional regions of proteins which have been called domains (Gilbert, 1978). Briefly, protein domains are discrete segments of a polypeptide chain that fold into a functional unit, such as a globular polypeptide segment that contains an enzyme active site. There is good evidence that immunoglobulin chains consist of such domains. The light (L) chains consist of two domains containing the variable (V) and constant (C) regions of the protein, while the basic heavy (H) chain structure consists of a single V domain linked to three C domains. The C region domains of a heavy chain are homologous to one another, and to the L chain C domain. The structure of the expressed immunoglobulin gene exactly mirrors these domains: L chain genes contain a major intron separating the bipartite V region from the C region;* and the constant region of the H chain genes consists of three domains separated by two intervening sequences in the case of the α heavy chain. Even more striking is the fact that an extra, short domain present in γ_1-type H chains, called the hinge region, is encoded by a fourth separate exon (Sakano *et al.*, 1979; Early *et al.*, 1979).

Unfortunately, the boundaries of the protein domains are not known with sufficient precision to test these ideas rigorously, but the general idea seems to apply to most genes for which the model has been tested. Although globin appeared at first sight not to have an obvious domain structure, recent work suggests that the central coding block corresponds to a haem-binding domain (Craik *et al.*, 1980), while the other domains are involved in α—β globin polypeptide chain interactions.

4 Evolution of split genes by exon-juggling

The fact that the domain-coding exons are clearly discrete units separated by "silent" DNA suggests a plausible model for gene evolution. It has been suggested (Gilbert, 1978) that split genes offer the attractive advantage that new gene combinations can be obtained by recombination between two genes within their introns (Fig. 14). Clearly such a mechanism can generate new combinations of genetic material with novel properties of potentially great selective advantage to the organism. Imagine, for example, recombining the exon for an active site of a given enzyme with the exon for a membrane-binding segment — conceivably the protein would be presented on the cell surface and by doing so, confer new properties on the cell. If this type of recombination occurred, we would expect to find homologous protein domains in diverse genes that otherwise have little homology. A persuasive example for this is again the

*The V region itself is encoded at two positions: V and J (or joining). The latter is closely linked to the C region on the chromosome. A discussion of immunoglobulin gene organization is beyond the scope of this article.

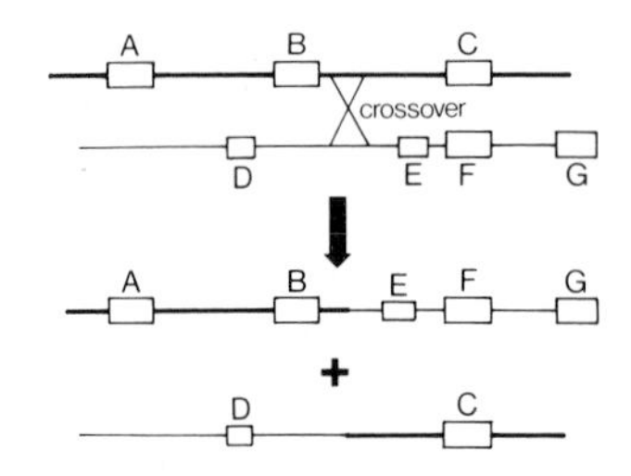

Figure 14 The generation of a new gene by exon-juggling. Recombination between two genes occurs at sites within introns. A reciprocal recombination generates two new genes which will potentially have new "hybrid" properties.

immunoglobulins. Sequences homologous to C region domains are present in both chains of HLA antigens — the small subunit or β_2-microglobulin and the heavy chain α-subunit (H. Ploegh, personal communication); it is noteworthy that otherwise little homology exists between these chains. A similar example is the complement component Clq which contains a protein segment related to collagen.

A second way that exon-juggling can create a new protein is by the internal amplification of exons within a gene. It seems possible that the three C regions of the immunoglobulin H chain evolved from the L chain gene by amplification of the single L-chain C region; alternatively the L chain may have evolved from the H chain by deletion of two exons.

C Linkage and gene organization

The relative location of genes on chromosomes is of interest for several reasons. First, there is reason to believe that the position of a gene on a chromosome influences its functioning. For example, expression of the human β-related globin genes (and presumably, therefore, that of many other genes) is influenced by the presence of cis-acting DNA sequences found within the gene cluster, but several thousand base pairs from the gene whose activity is modulated. In addition, transposition of a DNA segment to a different part of a chromosome affects the level of gene expression (position affects; see Spofford, 1976). Second, we would like to know how much chromosomal DNA has a coding function and how much is (dispensable?) spacer.

Detailed gene linkage studies were first carried out in the human β-related globin gene locus. The first determination of linkage at the DNA level for mammalian single copy genes was the construction of a physical map of the human δ- and β-globin genes using genomic blotting (Flavell *et al.*, 1978). The approach used was to construct physical maps of the DNA region surrounding these genes using restriction enzymes that generate fragments large enough to contain two genes. It was then possible to show that a given "two-gene" large fragment is cleaved once by a number of restriction endonucleases to give two single-gene fragments. In each case, the sizes of the two double-digest fragments adds up to the size of the original fragments (Fig. 15). Though this approach can be used to link up genes separated by up to 20 kb, if the intergenic distances are larger it is difficult to find restriction enzymes that only cleave once between the genes. To an extent this problem can be solved by using partial digests — then several cleavage sites for a given enzyme can be mapped. Clearly, if probes are available for both genes, then it is possible to predict the position of the cleavage sites from both genes. If the genes are linked the sites predicted will be coincidental when measured from both reference points (Fig. 15). This approach has also been used to link the human $^{A}\gamma$- and δ-globin genes (Bernards *et al.*, 1979).

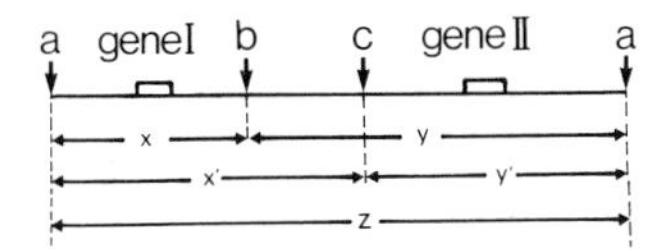

Figure 15 The demonstration of linkage of two genes by genomic blotting. Enzyme a cuts the chromosomal DNA to give a single fragment of length z. Double digests of enzyme a plus enzyme b or c gives two fragments in each case where $x + y = z$ and $x' + y' = z$. The positions of the cleavage by enzymes b and c can be verified by double digestion with other enzymes whose sites have already been mapped with respect to those for enzyme a.

Although it is possible to use genomic blotting to construct linkage maps of 40 kb or more (as we have done for the human $\gamma\delta\beta$ locus; Bernards *et al.*, 1979), in practice it is now more useful to construct the linkage map by isolating overlapping phage or cosmid clones. For example, Fig. 16 shows the structures of a series of overlapping phage λ clones containing the human β-related globin region which were isolated by Fritsch *et al.* (1980). To establish a linkage map for these clones, it is essential to show that the DNA segments overlap. To do this, two basic approaches are used:

(a) show that the clones have an identical restriction map in the overlapping region;
(b) isolate a fragment from an extremity of one of the clones, e.g. the left hand end of the first clone, and show that this hybridizes to the expected fragments of the right hand end of its overlapping clone.

We have isolated overlapping cosmids of the human β-related

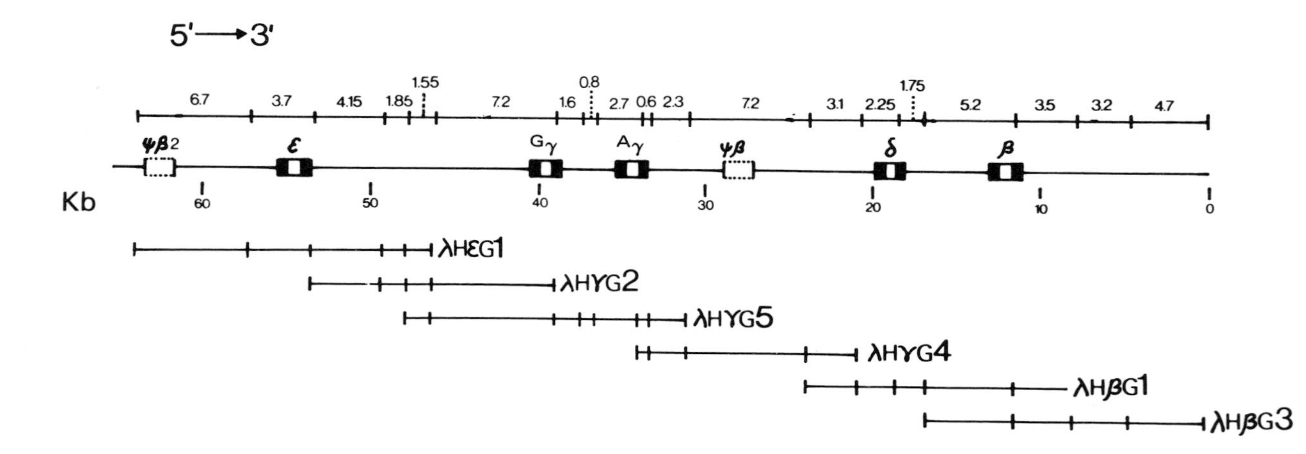

Figure 16 The linkage map of the human β-globin gene locus as shown by the structural analysis of overlapping phage λ clones. The genomic segments of the clones isolated are shown together with the cleavage sites for the enzyme Eco RI (redrawn from Fritsch *et al.*, 1980). The numbers indicate the size of fragments in kb.

globin gene region and these are shown in Fig. 17. As expected, it can be seen that fewer clones are required to cover an area of a given length than in the case of phage λ clones. In principle the availability of large clones should minimize some of the problems associated with "chromosome-walking" experiments (see below).

The gene clusters that have been best characterized to date are the human, rabbit and chicken β-related globin genes and the human and chicken α-related globin genes. The results obtained with the human and rabbit globin system can be summarized as follows.

1 The related genes are clustered in a short region of the chromosome

They are present in the order $5'\epsilon^G\gamma^A\gamma\delta\beta3'$ for the β-related human globin gene locus and $\zeta\alpha_1\alpha_2$ for the human α locus (see Maniatis *et al.*, 1980 for a review). In this case the genes are present on the chromosome in the order that they are expressed, but this does not appear to be so for the chicken globin genes, where the two adult α genes are flanked on either side by an embryonic α-globin gene. Though the α- and β-related genes are normally on separate chromosomes, this is clearly not obligatory since in *Xenopus* the α- and β-globin genes are closely linked (Jeffreys *et al.*, 1980). Within a given cluster the globin genes appear to be transcribed from the same DNA strand.

2 The distances between the globin gene

These by far exceed the size of the genes themselves. Thus, the distances are ϵ 13.5 kb $^G\gamma$ 3.5 kb $^A\gamma$ 13.5 kb δ 5.5 kb β; each globin gene (including transcribed introns) is about 1.5 kb in length.

3 Repetitive DNA within gene clusters

Little is known about the structures present in the intergenic DNA. Two types of DNA sequences have been recognized to date. First, numerous reiterated sequences are found at several sites within the β-related globin gene cluster. These have been mapped extensively in the case of the rabbit genes and Fig. 18 shows a summary of this work by Shen and Maniatis (1980). Though the position of these repeats is well characterized, nothing is known of their function. The characterization of repeated sequences in the human globin gene system is less advanced than in the rabbit. The so-called Alu family of repeats, however, have been the subject of considerable attention. About 300 000 of these sequences, each approximately 200 bp in

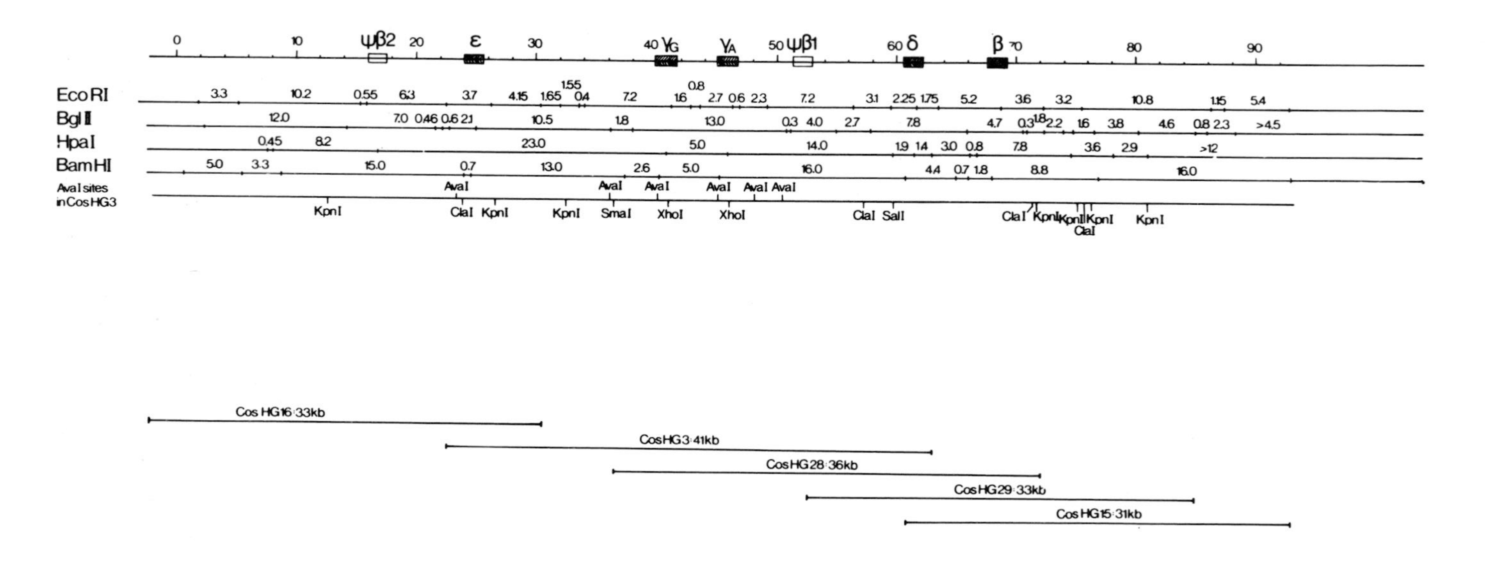

Figure 17 Overlapping cosmid clones containing the human β-globin gene locus. The overlapping clones were generated from human placental DNA partially digested with Mbo I and cloned in the vector pJB8 (Grosveld *et al.*, 1981) or p0PF1 (unpublished).

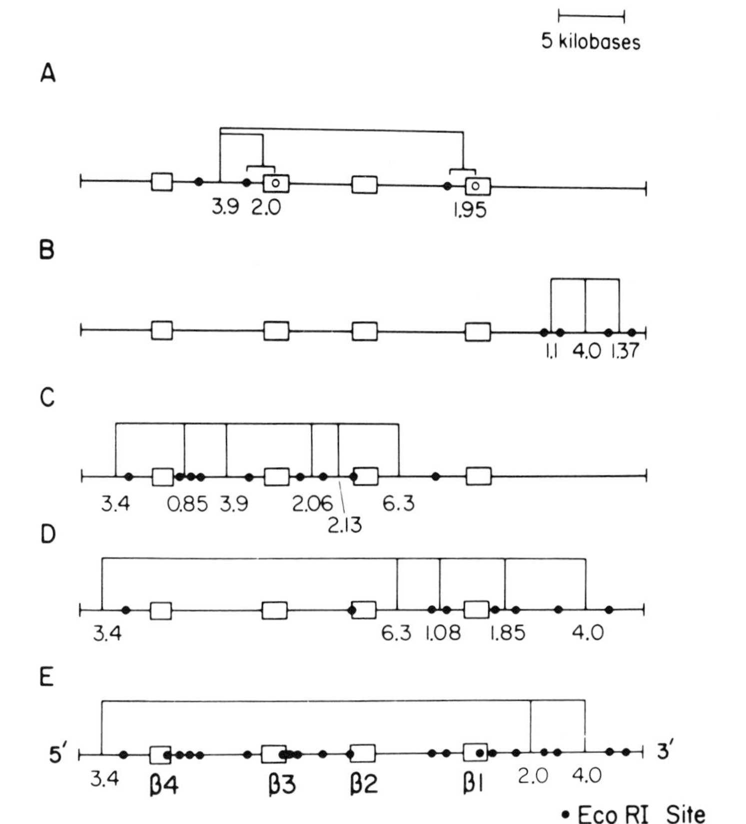

Figure 18 Classification of families of repeated DNA sequences in the rabbit β-globin gene cluster. Cross-hybridization experiments allow the resolution of 5 families of repeated sequences; their positions (A—E) are shown relative to the 4 rabbit β-globin genes (β2 is now known to be a pseudogene). From Shen and Maniatis (1980) with permission of MIT press.

length, are present in the human genome and they have a set of intriguing properties (Jelinek *et al.*, 1980):

— they are homologous both to an abundant class of small nuclear RNAs and with the double-stranded regions of heterogeneous nuclear RNA. This shows that at least some of the repeats, though not necessarily all, are transcribed *in vivo*.

— they serve as efficient transcription templates for RNA polymerase III *in vitro*

— they contain a sequence that is homologous to a sequence found near the replication origin of SV40, polyoma and BK viral DNAs.

From the last observation it is tempting to speculate that these sequences may represent cellular origins of replication. There is, however, no evidence to support this idea at present.

4 *Pseudogenes*

The availability of large cloned fragments from a given gene region also makes it possible to ask whether other mRNA coding sequences are linked to the genes under analysis. When short DNA fragments from the human β-globin gene regions were hybridized with probes for the globin gene coding sequences, additional fragments were detected as well as the known globin gene fragments. Since all the human β-related globins were apparently accounted for, the possibility arose that these globin gene-related sequences were defective genes which are not expressed and that had hitherto remained undetected. These genes have therefore been called pseudogenes in accordance with the terminology first used for a defective *Xenopus* 5S RNA gene which lacked its 3′ terminal sequences. This view has been confirmed in a number of cases by detailed DNA sequence analysis. Thus, there is a human α-related pseudogene (α_1) that is defective. It is 70—80% homologous to the adult α-globin gene throughout its sequence, but has a deletion of 20 nucleotides in the middle of the gene that puts the coding sequence out of phase (Proudfoot and Maniatis, 1980).

Several other pseudogenes have been characterized. Except for the fact that they are defective, they have few common structural similarities. Thus, the entire DNA sequence has been determined for a mouse pseudo-α-globin gene (Nishioka *et al.*, 1980). This gene lacks both introns and has several additional short deletions and insertions which serve to add stop codons in all reading frames. Pseudogenes have also been described in the β-related globin genes and at this stage it looks as if they will be a structural feature of all globin gene clusters and by extrapolation, all gene clusters. If we assume that the proportion of pseudogenes to active genes in other gene systems is similar to that for the globin genes, then it would follow that there are approximately equal numbers of pseudogenes and structural genes present in higher organisms.

Why are there pseudogenes? One view is that they represent the defective products of gene duplication and that they have no function. There is little evidence against such an idea. If this were so, then we would expect pseudogenes to be relatively recent in evolutionary terms; after longer times they would be lost by deletion. Alternatively, it could be argued that pseudogenes serve an important function. In this case, it is reasonable to expect them to be conserved throughout evolution. A detailed comparison of pseudogenes from related organisms should provide a solution to this problem.

5 *Gene-correction mechanisms*

Comparisons between the DNA sequences of a given gene in different but related organisms has shown that, in general, the intronic DNA sequences diverge more rapidly than the coding sequences. Surprising results have been obtained from a comparison of the structures of two given duplicated genes within the same organism. The two non-allelic γ-globin genes code for proteins that differ at only a single amino acid position; $^{G}\gamma$-globin has glycine and $^{A}\gamma$-globin has alanine at position 136. Normally one would expect the DNA sequence to show more differences since the third position of most codons can be varied without altering the amino acid encoded, and indeed, it is generally true that a nucleic acid sequence is more divergent than the protein sequence. The sequence of both the $^{G}\gamma$- and $^{A}\gamma$-globin genes has recently been reported by Slightom *et al.* (1980). A diploid human cell-line contains four γ-globin genes; two from each chromosome. Slightom *et al.* (1980) sequenced the $^{G}\gamma$ and $^{A}\gamma$ genes from one chromosome, together with allelic $^{A}\gamma$ gene from the second chromosome. The sequence of all these genes was strongly conserved and surprisingly this was also true for both introns. Amazingly, the sequences of the $^{G}\gamma$ and $^{A}\gamma$ genes on a single chromosome were more similar to one another than were the sequences of the two allelic $^{A}\gamma$ genes! A detailed study shows that the latter observation holds for the DNA sequences of the 5′ two-thirds of the large intron (the sequences further 5′ than this are almost identical in all three genes, so no comparison is possible). In the remainder of the large intron preconceptions apparently prevail, since here the allelic $^{A}\gamma$ genes are more similar to one another than are the $^{G}\gamma$ and $^{A}\gamma$ genes (Fig. 19). The DNA sequence at the boundary of these two dissimilar regions is a segment of simple sequence DNA consisting of alternating TG or CG and the authors have therefore suggested that this acts as a hot spot for recombination. They suggest that the sequences of the two non-allelic genes on a single chromosome are kept similar by recombination events, which serve to correct the sequence of the one gene against the other (Fig. 20). The mechanism proposed is an exchange between identical "sister" chromatids which are aligned during mitosis, although there are other possible explanations.

There is some evidence that this phenomenon may well be widespread. The two human non-allelic α-globin genes are more similar than expected (Efstratiadis *et al.*, 1980), despite the fact that the α gene duplication is found in most primates. In addition, Goossens *et al.* (1980) showed that certain humans have three α-globin genes on a single chromosome. This is what would be expected from an unequal crossing-over event and it is one of the specific predictions of the recombination mechanism detailed in Fig. 20.

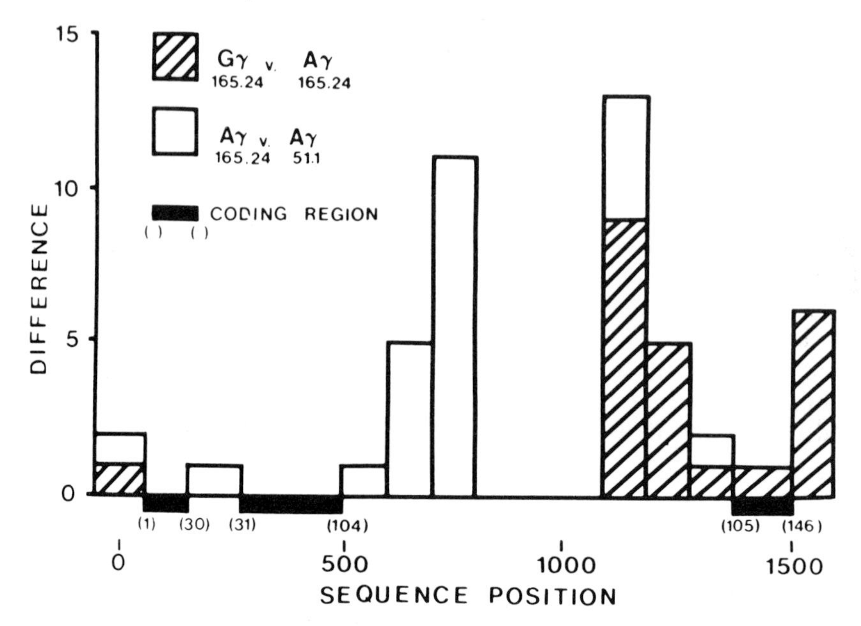

Figure 19 Bar diagram illustrating the distribution of differences between the non-allelic $^{G}\gamma$ and $^{A}\gamma$ genes (■) and between the allelic $^{A}\gamma$ genes (□). Base substitutions are counted as one difference; gaps, regardless of their length, are arbitrarily counted as three differences, so as not to over-emphasize gaps. Each bar shows the differences found in approximately 100 nucleotides, but the bars have been adjusted to coincide with the intron—exon boundaries. From Slightom *et al.* (1980) with permission of MIT press.

Unfortunately, to date we have little detailed information on the structure of the loci that contain other gene families. In most cases that have been studied, however, it is apparent that genes exist in families. Several related vitellogenin genes have been cloned from *Xenopus*, but their linkage has not yet been established (Wahli *et al.*, 1980). In other examples, there is good genetic evidence for the existence of gene clusters, but this has yet to be verified directly at the level of DNA. For example, the histocompatibility loci of man and mouse have been mapped as a tight cluster containing several structurally related genes (Bodmer, 1978).

D Chromosome-walking techniques

It is obviously of interest to ask: what lies next to a given gene cluster on a given chromosome? Current ideas are that gene families are located within a chromosomal domain (plausible but supported by little direct evidence!) and it would be of interest to define the boundaries of such a domain. To do this, a primary requisite is to

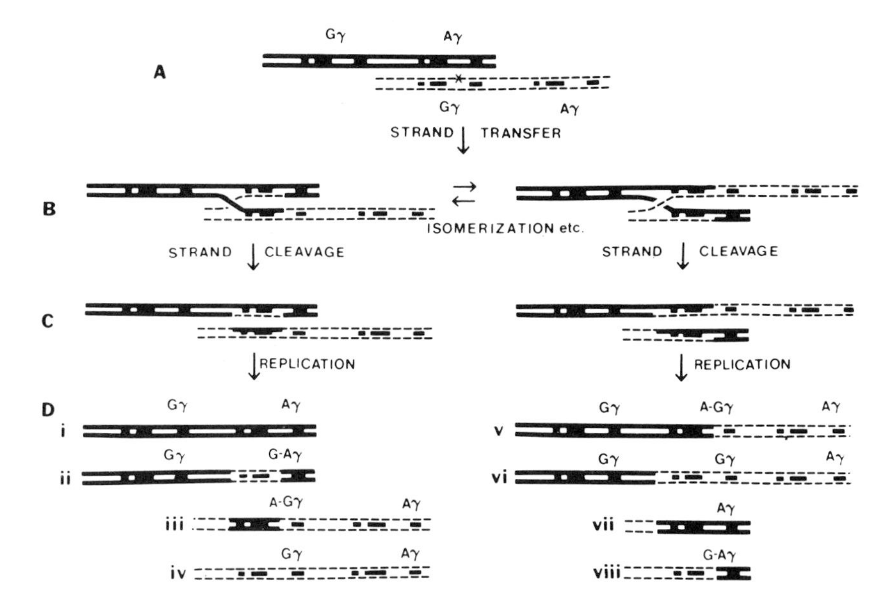

Figure 20 Diagrammatic representation and possible outcome of a recombinational event between a $^{G}\gamma$-globin gene aligned with an $^{A}\gamma$-globin gene. The participating DNA molecules are distinguished by light and heavy paired lines for illustrative purposes. The black bars between the paired lines indicate the globin coding regions. (A) The alignment at the time when the recombinational event was presumed to be initiated at one of the simple sequence hotspots (X) of IVS2. (B) Strand transfer, isomerization, branch migration and ligation yield the isomers. (C) Strand cleavage leads to the intermediate heteroduplex products. (D) DNA replication then leads to the final products. The products ii and iii show conversion (see text). From Slightom *et al.* (1980) with permission of MIT press.

isolate a series of overlapping clones upstream and downstream from a given cluster. The principle of the approach is outlined in Fig. 16. A given clone, say at the most 5′ extremity of the cluster, is fragmented by a restriction endonuclease to generate a terminal fragment containing only single copy DNA. This is then labelled *in vitro* by nick translation and hybridized to the phage or cosmid library that was used to obtain the original clones. This makes it possible to isolate more clones containing this 5′ region and to provide a second series of overlapping clones, some of which will extend still further upstream. The most 5′-terminal of these clones can then be reisolated and the procedure repeated.

This approach takes on a whole new significance when we consider the genome of *Drosophila melanogaster*. *Drosophila* is genetically the best characterized multicellular organism. Of particular value is the fact that several tissues contain polytene chromosomes which are large enough for detailed morphological studies; these chromosomes consist of a characteristic and (within a strain) invariant pattern of bands. The last few decades of genetic analysis have facilitated the painstaking localization of genes to bands. Though this localization is imprecise on molecular terms — a band can be several hundred kb — it is of great value for the identification of genes in general and particularly for those that are well defined genetically, but poorly in molecular terms. The major technical breakthrough was the application of *in situ* hybridization techniques to polytene chromosomes; it was shown that a labelled probe could by hybridized to these chromosomes and detected by autoradiography. Many of the genes of greatest interest to the biologist are difficult to analyse by the standard techniques of molecular biology. Mutations in these genes have a profound effect on the phenotype of the organism, yet we do not know anything about the proteins they encode — if indeed they serve such a function at all. For example, several so-called homeotic mutations have been described in *Drosophila* that apparently alter the cell determination stages during the development of the fly. For example, mutations in the bithorax locus (*bx*) cause aberrant development in the thoracic and abdominal regions. It is therefore believed that *bx* encodes a component that plays a crucial role in cell determination decisions in *Drosophila.*

The chromosomal position of *bx* is known but the molecular nature of its gene product is a mystery. Clearly a major step would be to clone the *bx* region. To do this, W. Bender and colleagues, in the laboratory of David Hogness, set up a protocol based on chromosome walking. As a starting point a clone is chosen that hybridizes with a band close to the gene *bx*. The walking procedure then commences. To facilitate rapid walking (jogging?), overlapping clones can be isolated from *Drosophila* mutants which had inverted a segment of the chromosome. This allows the isolation of a clone containing the junction fragment at the inversion. The DNA segment spanning the inversion, and within this clone, provides a probe that can be used to isolate clones on the distal side of the inversion from the original library from normal flies. The net effect is therefore to jump across the large distance covered by the inversion.

A problem with the chromosome-walking approach is the presence of repeated sequences at the end from which you intend to walk further. A repeated DNA sequence cannot be used to probe for its contiguous DNA sequences, since it will hybridize to many other clones from different regions of the genome in addition to its neighbouring DNA segments. Two solutions to this problem exist. It is either necessary to isolate a clone that spans the repeated sequence, yet which has single copy DNA to either side of the repeat. In this case, the single copy segment on the far side of the repeat can be used as a probe for further walking. If it is not possible to find a clone that spans the repeated segment in the library from the first fly strain, it is frequently possible to use the original single copy DNA segment to probe libraries from other fly strains. Commonly the homologous site in the second fly strain lacks the repeated sequence, so that more distal single copy DNA sequences can be isolated. These DNA sequences can then be used as probes for *in situ* hybridization to polytene chromosomes from the original fly strain to verify that they are indeed derived from the relevant region of the chromosome. Chromosome-walking experiments are likely to be widely used for the analysis of the genome of most higher organisms in the future.

E Transcriptional studies *in vivo*

It is of major importance to determine the mode of transcription of a gene. Although little foolproof evidence is available, it seems highly plausible that one of the primary ways that gene expression is regulated is at the level of transcription. An essential first step in the characterization of this process is to define the nature of the primary transcript of a given gene. This in turn points to those DNA sequences responsible for the production of the transcript, the promoter and regulatory sequences.

Before the availability of cloned DNA there was little progress in this field. From pioneering work, for example, of Curtis and Weissmann (1976), it was possible to deduce that pulse-labelled transcripts of a specific gene, in this case the mouse β-globin gene, were larger than the corresponding mRNA. The advent of cloning provided purified DNA from several genes and made possible the detailed characterization of primary transcripts.

As already discussed above, electron microscopy provides a powerful means of transcript mapping. By forming R loops between an mRNA and its gene, it is possible to show that a gene contains introns, and to count the minimal number of introns within a gene. By the same token, it is possible to test whether precursor mRNAs exist in the cell which contains both intronic and exonic RNA sequences. One of the earliest experiments to be performed with the mouse β-globin gene answered this point. Curtis and Weissmann had prepared large amounts of the purified mouse β-globin pre-mRNA, which is about 1500 nucleotides in length. Tilghman, Leder and their colleagues had recently cloned the mouse β-globin gene and shown that its size was of the same order as that of the pre-mRNA. To test whether the pre-mRNA was a transcript of both mRNA-coding and intron sequences, the two groups collaborated on an R looping project. If the pre-mRNA is a co-linear transcript of exons plus introns, a single continuous R loop will be formed between the pre-mRNA and its gene and this will map at the same position on the cloned DNA as the gene in question. This is what was found (Tilghman *et al.*, 1978a; Fig. 21). This result was consistent with the RNA processing model proposed for the generation of adenovirus mRNAs by a number of groups (Dunn and Hassell, 1977; Berget *et al.*, 1977; Chow *et al.*, 1977). In this well-known RNA splicing model, the entire gene is first transcribed and subsequently intron transcripts are removed by a sequential cleavage–splicing mechanism. If a gene contains several introns it is possible to use EM techniques to look at intermediates in the splicing process, for example to ask if there is a specific order in the removal of introns. This approach has been elegantly exploited by Chambon, O'Malley and their respective colleagues in the study of the splicing of chicken ovalbumin pre-mRNAs (e.g. Chambon *et al.*, 1979). This gene is considerably more complex than the globin genes, having seven introns and spanning some 8000 base pairs. R loops formed with nuclear RNA from chicken oviduct, the site of synthesis of ovalbumin, indeed show some molecules that lack some, but not all, of the introns (Fig. 22). By analysing the structure of a large number of partially spliced molecules, it is possible to ask how the introns are removed. From their studies it can be concluded that there is not a simple specific order of removal of intronic sequences from the ovalbumin pre-mRNA (Chambon *et al.*, 1979).

Two other techniques have been invaluable in the study of *in vivo* transcripts. First Berk and Sharp (1977) developed an elegant and specific method (Fig. 23) for the accurate determination of the termini of primary transcripts and the positions of the junctions of the introns and exons. The approach as applied in our laboratory (Grosveld *et al.*, 1981) to the analysis of rabbit β-globin pre-mRNAs is shown in Fig. 23. An RNA preparation which contains pre-mRNAs is hybridized with a denatured cloned DNA segment under high formamide conditions where DNA–DNA annealing is prevented. In the case of a primary transcript, the RNA forms hybrids which are co-linear with the gene, whereas in the case of spliced molecules, the hybrids formed contain DNA intron loops at the site where a splicing reaction has removed intron transcripts. The hybrids are then treated with S1 nuclease, which destroys single stranded DNA, but leaves duplex DNA and DNA–RNA hybrids intact. It follows, therefore, that a hybrid formed with a co-linear transcript will be reduced by S1 nuclease treatment to a perfect duplex of the same length as the RNA molecules; that length can be determined by gel electrophoresis

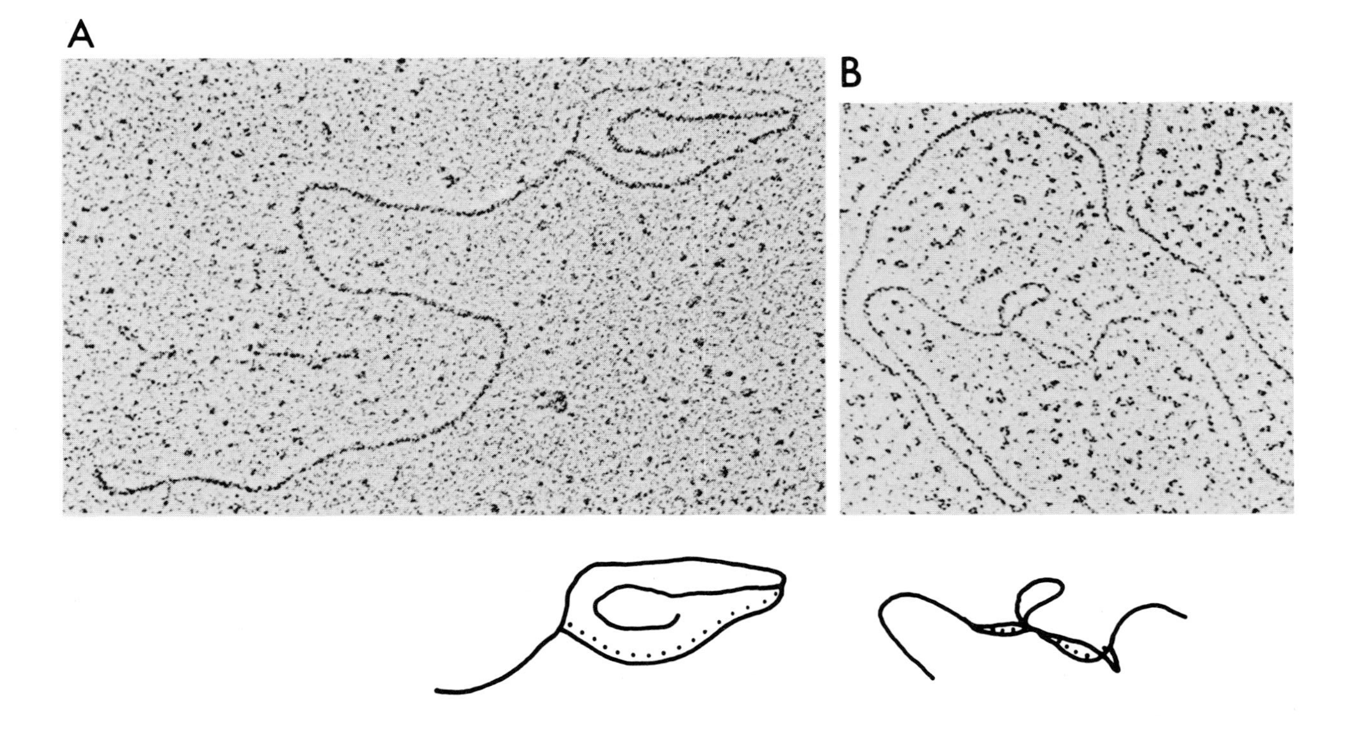

Figure 21 A. A co-linear R loop formed between the mouse β-major globin gene and 15S β-globin pre-mRNA. B. An R loop formed with 9S β-globin mRNA. From Tilghman *et al.*, (1978b).

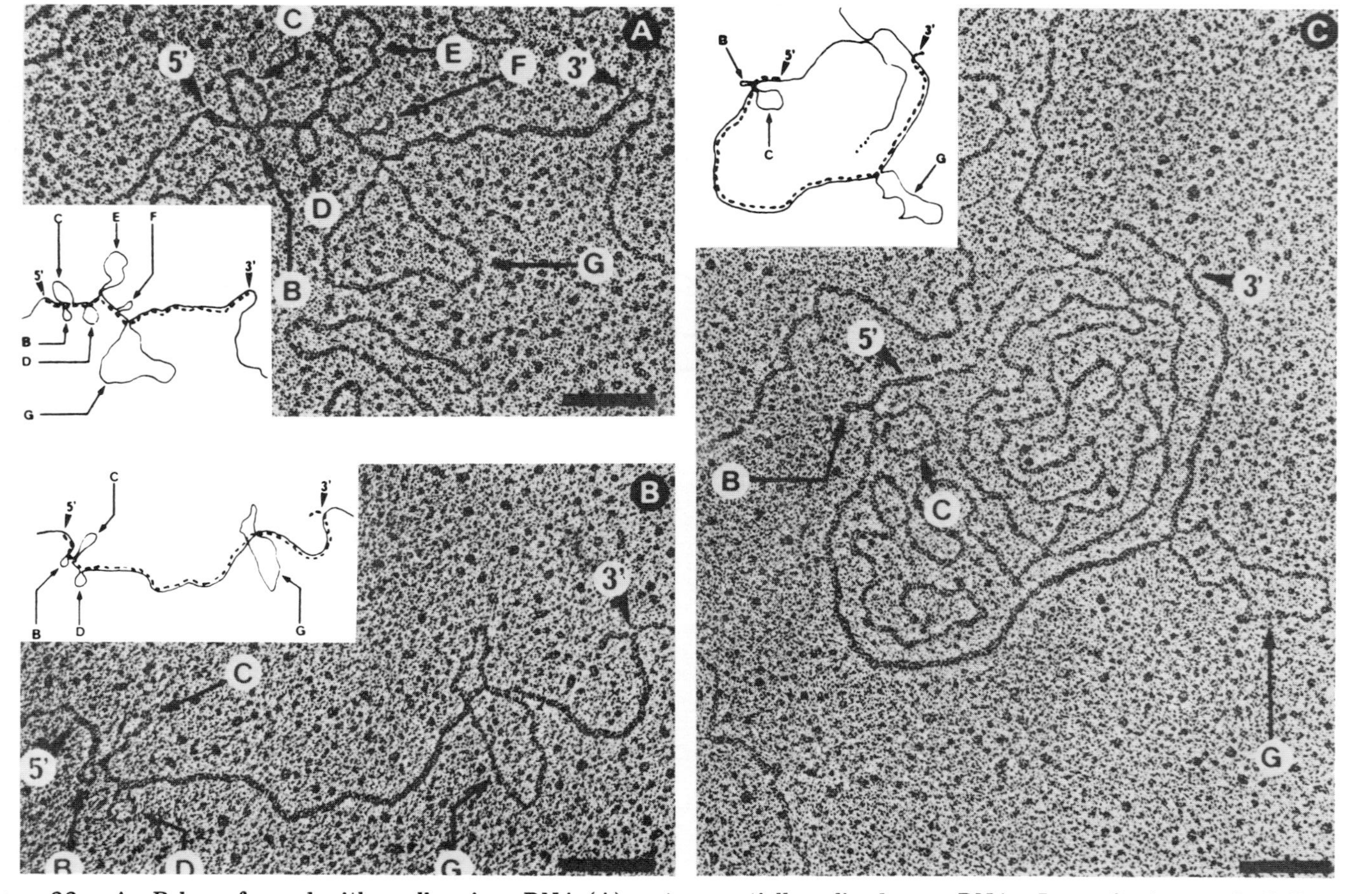

Figure 22 An R loop formed with ovalbumin mRNA (A) or two partially spliced pre-mRNAs. Loops for introns B, C, D and G can be seen in (B) and for B, C and G in (C). The remaining introns must therefore still be present in these pre-mRNAs. From Chambon *et al.* (1979).

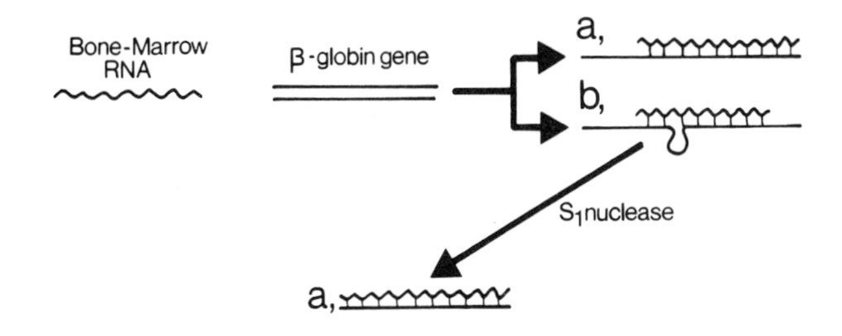

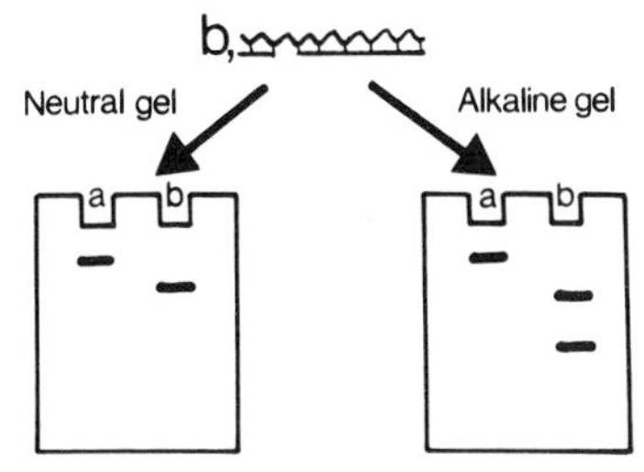

Figure 23 A scheme for mapping transcripts using S1 nuclease mapping. Rabbit bone marrow RNA is annealed with a cloned β-globin gene under high formamide conditions. The hybrids are treated with S1 nuclease and analysed by agarose gel electrophoresis at neutral and/or alkaline pH (see text for details).

at neutral pH. A hybrid formed with a spliced intermediate will also be reduced to a hybrid of the same length as the spliced RNA, but containing a nick in the DNA molecule at the position where the DNA loop was removed by the nuclease. The two types of molecules can be distinguished by gel electrophoresis under denaturing conditions, where the hybrid structure is disrupted; in the case of a co-linear transcript, a single fragment is seen of the same length as the hybrid, whereas a spliced transcript shows several fragments, each complementary to a segment of the RNA that has been spliced (Fig. 23).

The S1 nuclease-treated DNA fragments are most commonly characterized by using unlabelled DNA fragments in the hybridization reaction to RNA and by subsequently detecting the fragments that are rendered resistant to S1 nuclease by Southern blotting followed by hybridization to DNA probes for the same gene. A commonly used modification of this approach is to use terminally labelled DNA probes for hybridization with the pre-mRNAs followed by S1 nuclease treatment. If a sufficiently short fragment is used, it can be analysed on high resolution polyacrylamide gels that are capable of separating DNA molecules differing in length by only a single nucleotide. As a marker, the same DNA fragment is treated with standard

DNA sequencing reagents to provide a ladder which indicates each nucleotide of the sequence in the area of interest. By aligning the S1 nuclease-resistant fragment with the fragments in the "sequence" lanes, it is possible to read off the exact position at which the RNA protects the DNA against S1 nuclease, and thus the precise map position of the transcript within one or two nucleotides. Fig. 24 shows an example where this method is used to map the co-linear transcript of the rabbit β-globin gene. The 5′ end is mapped using a 5′ labelled DNA fragment overlapping the mRNA cap site and the 3′ end is mapped with a 3′-labelled fragment which overlaps the region containing the site at which polyadenylation occurs. From these, and many other experiments using similar approaches, it can be concluded that the 5′ and 3′ ends of discrete nuclear pre-mRNAs are the same as those of the mRNA: such pre-mRNA molecules are transcripts of mRNA coding plus intron sequences.

Another invaluable method that is used for transcription mapping is so-called "Northern" blotting (Alwine *et al.*, 1977). In this method, RNA is electrophoresed in denaturing agarose gels and then transferred by blotting to a chemically activated paper. On contacting the paper the RNA becomes covalently bound. After the remaining binding sites on the paper have been inactivated, the paper carrying the RNA can be hybridized with a cloned fragment of a gene such as ovalbumin. This method can be very useful since it is possible to hybridize with each intron separately to sort out the structure of the spliced intermediates. Fig. 25 shows an example of this method applied to the ovalbumin gene.

The results derived in a number of systems using some or all of the above approaches can be summarized as follows:

Figure 24 High resolution S1 nuclease mapping of the termini of pre-mRNAs. The 5′ and 3′ ends are mapped using short DNA fragments that are terminally labelled at their 5′ and 3′ ends (*) respectively; the fragments used overlap the respective ends of the pre-mRNA; S1 nuclease digestion of the hybrid generates a short DNA fragment of a length from the nucleotide carrying the end-label to the relevant terminus. Its precise length is determined by co-electrophoresis with the same fragment after treatment with the reagents that cleave at G + A or C + T sequence. These can be used to align the S1 nuclease resistant fragment with the known DNA sequence. The relevant parts of the sequence of the non-coding strand are shown; to read this sequence, lanes 4 (left side) and 1 (right side) should be read as C + T and lane 2 (right side) should be read as G + A. This is somewhat difficult to resolve at this manification. The relative locations of the terminal label, the 5′ and 3′ ends of the gene and the sizes of the S1 nuclease resistant fragments are shown. The asterisks indicate the positions of the labelled nucleotides. From Grosveld *et al.* (1981), with permission of MIT press.

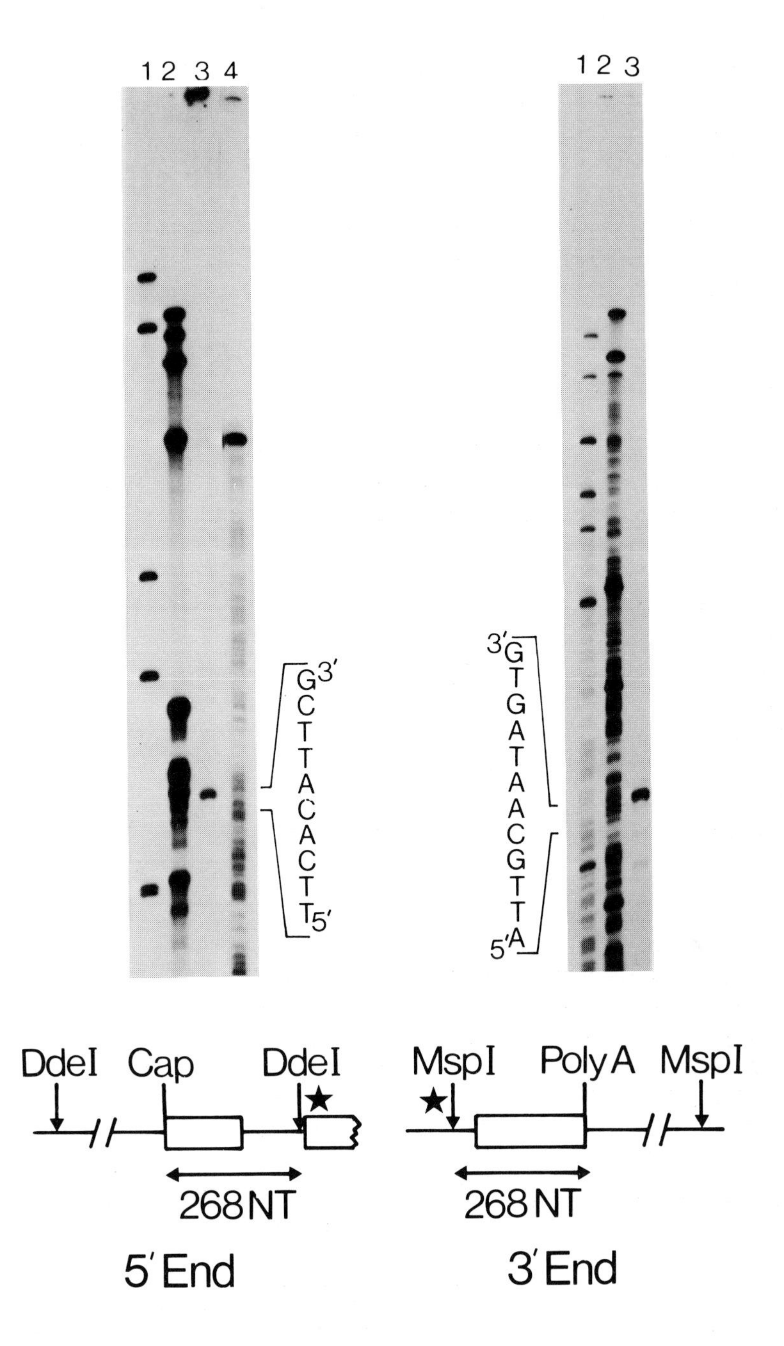

1 2 3 4
G 3′
C
T
T
A
C
A
C
T
T 5′
1 2 3
3′ G
T
G
A
T
A
A
C
G
T
T
5′ A
DdeI
Cap
DdeI
268 NT
5′ End
MspI
Poly A
MspI
268 NT
3′ End

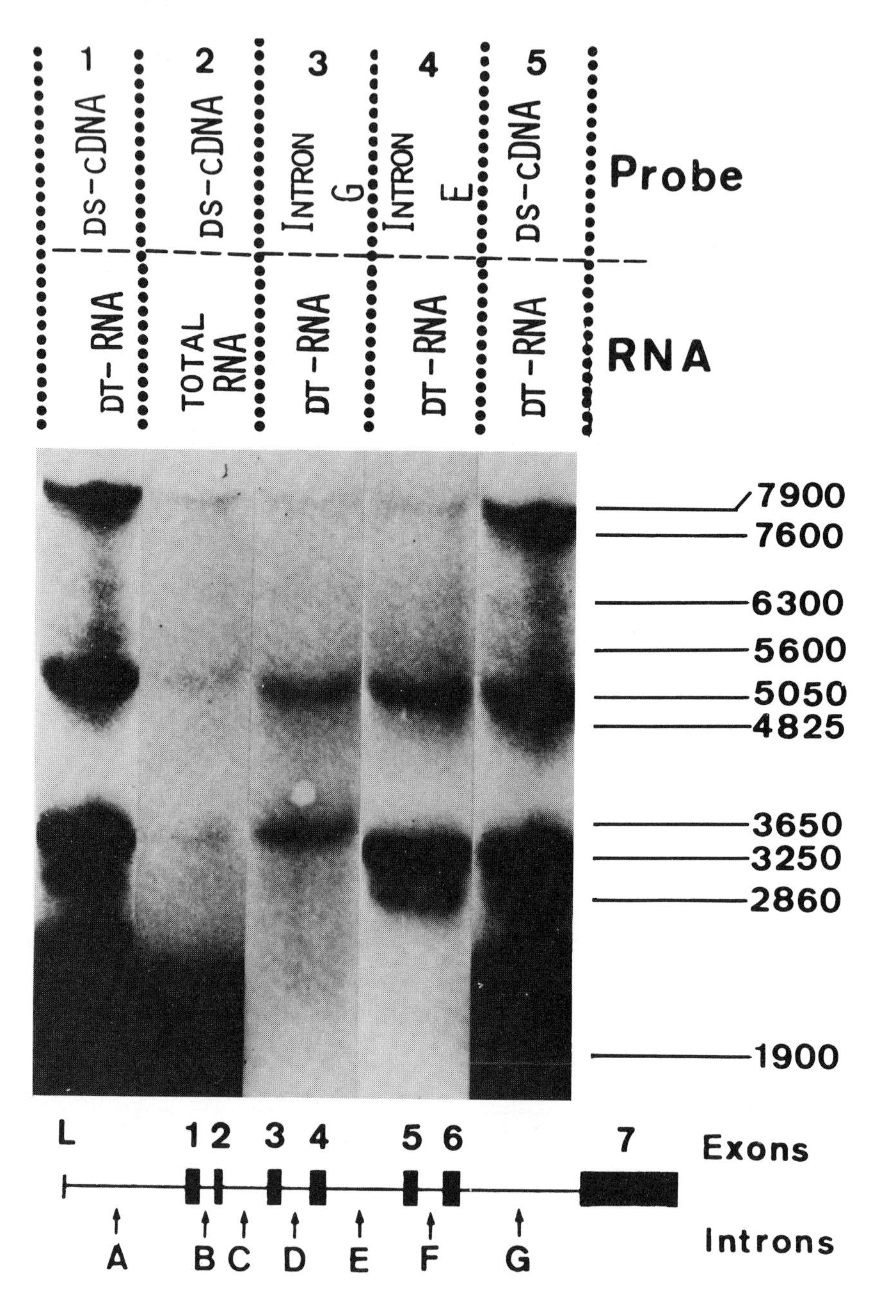

Figure 25 Analysis of pre-mRNA by "Northern" blotting. RNA from chick oviduct is hybridized with various probes from the chicken ovalbumin gene. The ovalbumin gene structure is shown schematically under the figure. DT RNA is poly(A)$^{+}$-RNA that has been obtained by oligo(dT)-cellulose chromatography. From Chambon *et al.* (1979).

(1) The 5′ and 3′ ends of the largest stable polyadenylated transcripts are co-terminal with the corresponding ends of the mRNAs (mouse β-globin (Weaver and Weissmann, 1979); rabbit β-globin (Grosveld *et al.*, 1981); chicken ovalbumin (Roop *et al.*, 1980)).
(2) Removal of each intron may occur in a complex fashion: a given intron can be removed in several steps, removing only part of an intron at any time (e.g. mouse β-globin (Kinniburgh and Ross, 1979); rabbit β-globin (Grosveld *et al.*, 1981)).
(3) The order of removal of introns is complex and in many cases unresolved; in some genes there is evidence for a preferred order of excision of certain introns (e.g. Grosveld *et al.*, 1981). Whether this has any profound significance, or simply reflects greatly different rates of excision of different introns during an otherwise random process is not known.

1 Regulation of gene expression by differential splicing?

RNA splicing makes possible the multiple use of a single given segment of genetic material. By splicing the transcript of this segment to alternative acceptor RNA segments, it is clear that hybrid mRNAs can be produced that encode hybrid proteins (Fig. 26). This type of mechanism is used in the early gene products of papovaviruses. In SV40 this region contains two overlapping genes which encode two tumour antigens, T and t: the mRNAs for each respective protein are obtained by different modes of splicing of essentially the same RNA. Since in both cases the first exon contains a large section of coding sequences, it follows that both proteins share a common *N*-terminal sequence. The *C*-terminal regions differ, however, since in T mRNA the stop codon of t is removed by splicing to expose a long protein-coding segment in the correct reading frames (Reddy *et al.*, 1979).

Multiple use of genetic information also occurs in adenovirus, where common leader sequences (that is, an exon which does not normally contain protein-coding mRNA) occur (Chow *et al.*, 1977). Indeed, it is generally argued that this type of mechanism is ideally suited to viruses. Viruses need to condense as much genetic information into as small a genome as possible, for smaller genomes replicate faster, and this confers an obvious competitive advantage. They are also smaller targets for cellular defence mechanisms.

Surprisingly, however, at least one cellular gene system seems to use this mechanism. α-Amylase is expressed mainly in three tissues in mammals: in the salivary gland, where it is secreted in the saliva; in the pancreas where it is secreted to the gut; and in the liver where

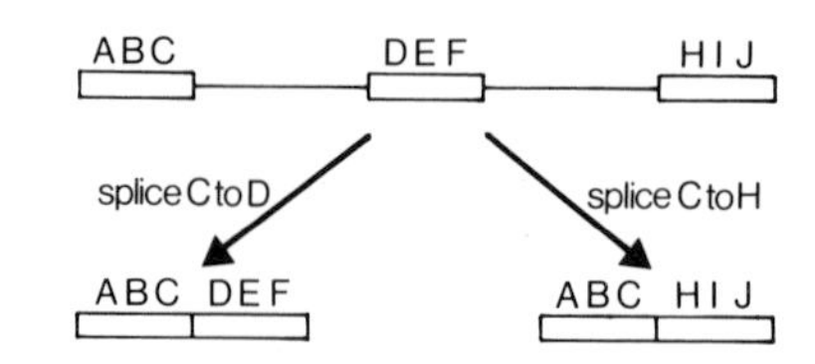

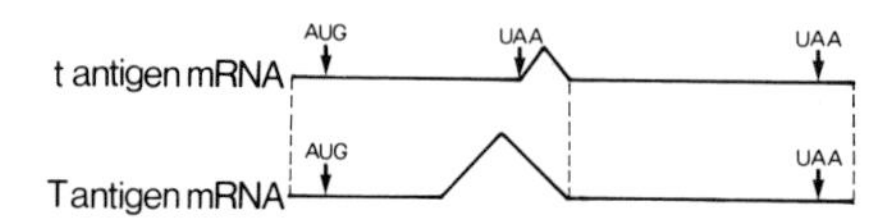

Figure 26 Multiple use of genetic information by differential splicing. In the top part of the figure the principle is illustrated. By splicing a given 5′ exon to two different 3′ exons, mRNAs are generated that encode proteins with part common and part different polypeptide sequences. In the bottom of the figure, the example of the early gene products of SV40 are shown. The pre mRNA contains two relevant termination codons. t mRNA is produced by splicing out a small segment immediately after the first termination codon; consequently the t polypeptide is produced by reading the mRNA from the AUG to this termination codon. In T mRNA, the first termination codon is eliminated by splicing and the polypeptide produced consequently reads to the second termination codon.

it has a role in glycogen degradation. In each mouse tissue a different α-amylase mRNA is found (Schibler *et al.*, 1980). Recent elegant studies (Young *et al.*, 1981) have shown that there are only two α-amylase genes to provide the three different α-amylases: the pancreatic α-amylase mRNA is encoded by one of these genes; the liver and salivary gland mRNAs are, however, both encoded by the same gene and are generated by an alternative splicing mechanism schematically shown in Fig. 27. The liver mRNA is coded by the second, third and fourth exons,* while the salivary gland mRNA is encoded by the first, third and fourth exons. Why does the mouse go to all this trouble? One hint comes from the levels of α-amylase mRNA in the liver and salivary gland. In the former case, there are about 100 molecules per cell, whereas in the latter case, 10^4 mRNA molecules are present per cell. Perhaps the transcription, processing or both, reactions required to produce this mRNA are more efficient for the

*It is important to note that this refers only to the 5′ portion of the mRNA. It is likely that there are several more exons — these are assumed to be the same for both mRNAs since the coding sequences are the same. This remains, however, to be verified.

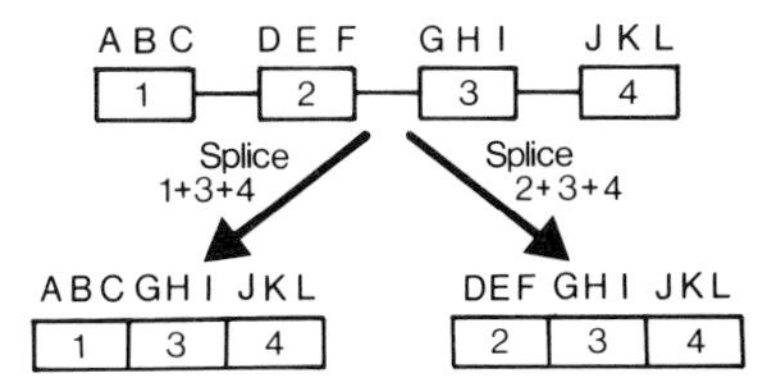

Figure 27 A schematic illustration of the differential splicing events in the mouse α-amylase genes. In the salivary gland the left-hand splicing scheme is followed; in the liver the right-hand scheme.

salivary gland mRNA. Sequence studies of the liver and salivary gland mRNAs show that they differ in that the liver mRNA has a 5′ untranslated region that is about 80 nucleotides longer than that of the salivary gland mRNA. Since the 5′ untranslated region is thought to be involved in ribosome binding, it is also possible that the efficiency of ribosome binding and therefore of translation is different for these two mRNAs.

2 *DNA sequences with a putative role in transcription?*

Comparison of the DNA sequence of a number of eukaryotic genes has pointed to DNA sequences conserved in evolution. First, an A—T rich sequence is present about 30 nucleotides in front of most, but not all, genes; the canonical sequence is TATAA (Gannon *et al.*, 1979). The similarity of this sequence with the Pribnow box segment of the *E. coli* promoter (canonical sequence TATPuATG) is suggestive of a role in transcription. About 80 nucleotides upstream from most eukaryotic genes the "CCAAT box", another conserved DNA sequence, is found (Efstratiadis *et al.*, 1980). Since the *E. coli* promoter has two functional regions around − 7 and − 35 (where + 1 is first nucleotide of the mRNA), it is not unreasonable to postulate that the CCAAT box may also be involved in initiation of transcription.

Evidence supports the role of the TATAA box in transcription but, as yet, it is not established whether the CCAAT box has a similar function. In addition, evidence is accumulating for a role of DNA sequences in the region of −200 to −1300; detailed sequence comparisons have yet to be made in this region for several genes and it is therefore too early to tell whether these are conserved DNA sequences at this position.

F "Reversed genetics" — Site-directed mutagenesis

Despite great advances in our understanding of the prokaryotic genes and their expression, we know little about the mechanisms of gene expression in eukaryotes. As is clear from the preceding discussion, much structural information on eukaryotic genes has been accumulated. In order for us to be able to understand how function co-ordinates with structure, we have to look at gene expression as well as structure.

The elucidation of the relationship between structure and function of the genetic material is greatly aided by the availability of suitable mutants. In classical genetics, mutants were obtained by first selecting a given phenotype, and second, examining the genetic alteration. Although this approach has been fruitful for prokaryotes, it is extremely difficult to screen for infrequent mutations in higher organisms. The mammalian genome is 1000-fold more complex than that of bacteria; most animals and plants have long generation times and experiments in classical genetics must necessarily consume much time; these organisms cannot be handled in useful numbers (the idea of 10^6 mice is nightmare-ish!) and they are usually diploid, which hinders the isolation of recessive mutations.

To solve this impasse, a procedure known as reversed genetics has been developed over the last six years (Weissmann, 1978). In this approach alterations are introduced into specific regions of DNA either by point mutation or by deletion or insertion. These mutated molecules can be purified by cloning in bacteria and the alterations in the clones characterized by DNA sequencing. A crucial second step in this approach is to ask what effect the mutation has on gene function. To do this, the mutated gene has to be tested for activity in a system, *in vivo* or *in vitro*, that expresses the normal gene in as physiological a manner as possible.

1 Methods

To generate specific mutations three approaches are followed: deletion or insertion of DNA sequences, or replacement of one or more bases by point mutation.

Deletion mutants (Fig. 28) can be prepared after cleaving the DNA with restriction endonucleases. For example, simple deletions may be obtained by cleaving at two sites within a circular plasmid DNA molecule, followed by recycling the plasmid and recloning. Fine deletions are obtained from a given point by the use of exonucleases on a defined DNA fragment. The fragment is first purified and digested with an exonuclease, such as exonuclease III, which degrades DNA in the $3' \rightarrow 5'$ direction. When a preparation is obtained with approximately the desired size of deletion, the single stranded regions of the second DNA strand are removed by S1 nuclease digestion followed by incubation with DNA polymerase I in the presence of deoxynucleoside triphosphates to render the ends blunt.

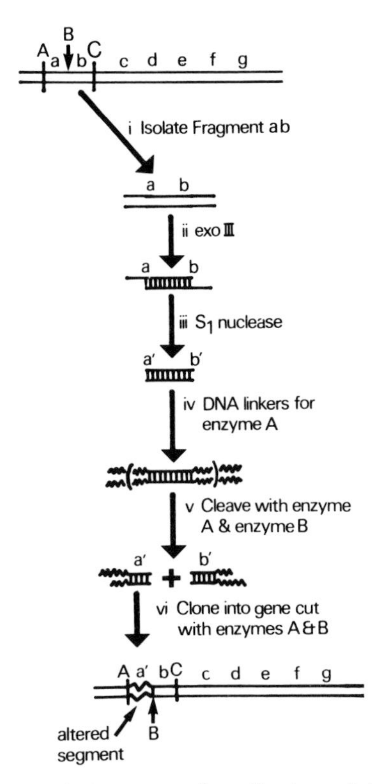

Figure 28 An example of the type of method used to generate specific deletions in a given DNA region.

The "mutated" DNA molecule must then be inserted into the gene from which it was derived. Usually this is done by the addition, by end ligation, of synthetic linkers which contain a useful restriction site. The mutated DNA fragment will usually be deleted from both ends of the DNA molecule. If deletions from only one end are required, then the DNA segment is cleaved in the central regions of the molecule to separate the two segments. The desired segment is then reinserted into the gene and cloned in *E. coli.* In this way a number of clones will be generated, each lacking different discrete segments of the gene region under investigation. DNA sequence analysis is used to determine precisely the end points of the deletions.

In order to introduce point mutations, it is necessary either to alter a base that is present in the DNA or alternatively, to introduce an abnormal base at the same position. Various methods can be used to alter bases *in situ*. For example, DNA can be treated with hydroxylamine which causes the modification of dC residues to N^4-OHdC. Since the latter has an altered tautomeric equilibrium such that it can base pair with both A and G residues during DNA replication, a significant fraction of the nucleotides inserted opposite this site during replication will be dAs: a point mutation results. Alternatively, DNA can be treated with bisulphite. This causes deamination of dC residues to dU and is therefore also mutagenic. Since this reaction is specific for single stranded DNA, it can be used to modify a specific restriction site at a single strand overhanging end (Shortle and Nathans, 1978).

To insert new "mutant" bases into a segment of DNA *in vitro*, DNA synthetic reactions are performed. A short specific single stranded DNA segment is used as a primer for synthesis of DNA in the region to be mutated. The synthesis reaction is carried out with up to three deoxynucleoside triphosphates to limit DNA synthesis to a short specific region under analysis. One of the normal deoxynucleoside triphosphates (e.g. dCTP) is replaced by a mutagenic triphosphate (e.g. N^4-OHdCTP). After insertion of the mutagen in this specific site, the mutagenic analogue is removed and the synthesis of the new DNA chain completed by the addition of all 4 dNTPs. The DNA is then cloned and the nucleotide sequence of the mutated region determined (for a general review, see Weissmann, 1978). An alternative to the use of nucleotide analogues is to use synthetic primers that are homologous to the region of interest, but that have one or more sequence differences. The primer is extended by DNA polymerase I with 4 dNTPs and then cloned. Again, sequence analysis is used to prove that the requisite mutation is obtained.

2 *Assay system for eukaryotic gene expression*

When a given gene and its putative regulatory mutants are available, it remains to ask whether these mutations alter expression of the gene. The gene is therefore either reintroduced into a living cell which is permissive for its expression, or added to a suitable *in vitro* system, where its transcription can be assayed directly (Wickens and Laskey, this series, Vol. 1).

The *in vivo* systems currently in use are:

(a) thymidine kinase$^-$ (tk$^-$) mouse L cells — in this case the gene in question is linked to the herpes (HSV-1) tk gene (Mantei *et al.*, 1979) or simply mixed with the tk gene (Wold *et al.*, 1979): tk$^+$ transformants are then selected. These transformants usually contain the gene under study and express this at a low level; RNA splicing apparently occurs efficiently (Mantei *et al.*, 1979; Lai *et al.*, 1980; Breathnach *et al.*, 1980). Similar transformation systems are available for yeast, where yeast genes that confer a selective advantage are used to obtain the transformants.

(b) The gene can be linked to a virus genome such as SV40 DNA. In this situation the mouse β-globin gene has been shown to be expressed under control of its own promoter.

(c) The gene can be introduced by microinjection into *Xenopus* oocytes (Mertz and Gurdon, 1977; Kressmann *et al.*, 1977) or animal cells. In *Xenopus* oocytes the genes appear to be transcribed faithfully and this system has been most useful for the analysis of the expression of the sea urchin histone genes.

Alternatively, transcription experiments can be performed *in vitro*. This works efficiently for genes transcribed by RNA polymerase III and more recently has been shown to work with RNA polymerase II, when either whole cell extracts are used (Manley *et al.*, 1980) or when 100 000 g cell supernatants are used to supplement purified RNA polymerase II (Weil *et al.*, 1979).

Experiments performed with these *in vivo* and *in vitro* systems have implicated the Goldberg-Hogness box, the CCAAT box and regions still further upstream in the initiation of RNA synthesis on eukaryotic gene templates by RNA polymerase II. This work is considered in detail by the article of Wickens and Laskey (this series, Vol. 1).

X Conclusions

It is evident from the preceding discussions that the cloning of genomic DNA in segments up to 40 kb has become essentially a routine technique in the last four years. It is possible to use these clones to analyse the structure, evolution and mode of expression of the genes isolated. With the availability of the techniques of reverse genetics, it is possible to test the function of any DNA sequence for which we have a sensitive assay. Thus eukaryotic promoters are being mapped in detail using *in vivo* and *in vitro* approaches. To date, we do not have adequate assay systems to study the developmental aspects of gene expression. All *in vivo* and *in vitro* systems employed seem to be permissive to expression and not to show even the most rudimentary aspects of cell-specific regulation. Relatively short DNA segments containing oviduct (ovalbumin) or erythroid (globin) genes are expressed in cells that never express them *in vivo* as far as we can tell. A crucial step for the future will, therefore, be the establishment of assay systems for developmental stage-specific expression.

XI References

Aaij, C. and Borst, P. (1972). The gel electrophoresis of DNA. *Biochim. Biophys. Acta* **269**, 192–200.

Alwine, J. C., Kemp, D. J. and Stark, G. R. (1977). Method for detection of specific RNAs in agarose gels by transfer to diazobenzyloxymethyl paper and hybridization with DNA probes. *Proc. Natn. Acad. Sci. U.S.A.* **74**, 5350–5354.

Armstrong, K. A., Hershfield, V. and Helinski, D. R. (1977). Gene cloning and containment properties of plasmid ColEl as its derivatives. *Science* **196**, 172–175.

Bacchetti, S. and Graham, F. (1977). Transfer of the gene for thymidine kinase to thymidine kinase-deficient human cells by purified herpes simplex viral DNA. *Proc. Natn. Acad. Sci. U.S.A.* **74**, 1590–1594.

Becker, A. and Gold, M. (1975). Isolation of the bacteriophage lambda A-gene protein. *Proc. Natn. Acad. Sci. U.S.A.* **72**, 581–586.

Beggs, J., Van Den Berg, J., Van Ooyen, A. and Weissmann, C. (1980). Abnormal expression of chromosomal rabbit β-globin gene in *Saccharomyces cerevisiae. Nature, Lond.* **283**, 835–840.

Bell, G. I., Piztet, R. L., Rutter, W., Cordell, B., Tischer, E. and Goodman, H. M. (1980). The nucleotide sequence of the human insulin gene. *Nature, Lond.* **284**, 26–32.

Benton, W. and Davis, R. (1977). Screening λgt recombinant clones by hybridization to single plaques *in situ. Science* **196**, 180–182.

Benzer, S. (1962). Fine structure of a gene. *Sci. Am.* **206**, 70–84.

Berget, S. M., Moore, C. and Sharp, P. A. (1977). Spliced segments at the 5′ terminus of adenovirus 2 mRNA. *Proc. Natn. Acad. Sci. U.S.A.* **74**, 3171–3175.

Berk, A. J. and Sharp, P. A. (1977). Sizing and mapping of early adenovirus mRNAs by gel electrophoresis of S1 endonuclease-digested hybrids. *Cell* **12**, 721–732.

Bernards, R., Little, P. F. R., Annison, G., Williamson, R. and Flavell, R. A. (1979). Structure of the human ${}^{G}\gamma\,{}^{A}\gamma\,\delta\beta$ globin gene locus. *Proc. Natn. Acad. Sci. U.S.A.* **76**, 4827–4831.

Bernhard, K., Schrempl, H. and Gorbel, W. (1978). Bacteriocin and antibiotic resistance plasmids in *Bacillus cereus* and *Bacillus subtilis. J. Bacteriol.* **133**, 897–903.

Birnsteil, M. L., Sells, B. H. and Purdom, I. (1972). Kinetic complexity of RNA molecules. *J. Mol. Biol.* **63**, 21–39.

Blattner, F. R., Blechl, A. E., Denniston-Thompson, K., Faber, H. E., Richards, J. E., Slightom, J. L., Tucker, R. W. and Smithies, O. (1978). Cloning human fetal γ globin and mouse α-type globin DNA: preparation and screening of shotgun collections. *Science* **202**, 1279–1284.

Bodmer, W. (1978). The HLA system. *Br. Med. Bull.* **34** (3).

Bolivar, F. (1978). Construction and characterization of new cloning vehicles. III. Derivatives of plasmid pBR322 carrying unique EcoRI sites for selection of EcoRI generated recombinant DNA molecules. *Gene* **4**, 121–136.

Bolivar, F., Rodrigues, R. L., Betlach, M. C. and Boyer, H. W. (1977a). Construction and characterization of new cloning vehicles. I. Ampicillin-resistant derivatives of the plasmid pMB9. *Gene* **2**, 75–93.

Bolivar, F., Rodigues, R. L., Greene, P. J., Betlach, M. C., Heynecker, H. L., Boyer, H. W., Crosa, J. M. and Falkow, S. (1977b). Construction and characterisation of new cloning vehicles. II. A multipurpose cloning system. *Gene* **2**, 93–113.

Bollum, F. J. (1962). Oligodeoxyribonucleotide-primed reactions catalysed by calf thymus polymerase. *J. Biol. Chem.* **237**, 1945–1949.

Breathnach, R., Mandel, J. L. and Chambon, P. (1977). Ovalbumin gene is split in chicken DNA. *Nature, Lond.* **270**, 314–319.

Breathnach, R., Benoist, C., O'Hare, K., Gannon, F. and Chambon, P. (1978). Ovalbumin gene: evidence for a leader sequence in mRNA and DNA sequences at the exon–intron boundaries. *Proc. Natn. Acad. Sci. U.S.A.* **75**, 1853–1857.
Breathnach, R., Mantei, N. and Chambon, P. (1980). Correct splicing of a chicken ovalbumin gene transcript in mouse L cells. *Proc. Natn. Acad. Sci. U.S.A.* **77**, 740–744.
Broome, S. and Gilbert, W. (1978). Immunological screening method to detect specific translation products. *Proc. Natn. Acad. Sci. U.S.A.* **75**, 2746–2749.
Cami, B. and Kourilsky, P. (1978). Screening of cloned recombinant DNA in bacteria by *in situ* colony hybridization. *Nucl. Acids Res.* **5**, 2381–2390.
Carbon, J., Ratzkin, B., Clarke, L. and Richardson, D. (1977). *In* "Molecular Cloning of Recombinant DNA", Miami Winter Symposia (Eds W. A. Scott and R. Werner) Vol. 13, 59–72. Academic Press, New York.
Carbon, J., Clarke, L., Chinault, C., Ratzkin, B. and Walz, A. (1978). *In* "Biochemistry and Genetics of Yeasts: Pure and Applied Aspects" (Eds M. Bacila, B. L. Horecker and A. O. M. Stoppani) 428–437. Academic Press, New York.
Casey, J. and Davidson, N. (1977). Rates of formation and thermal stabilities of RNA: DNA and DNA: DNA duplexes at high concentrations of formamide. *Nucl. Acids Res.* **4**, 1539–1552.
Catterall, J. F., O'Malley, B. W., Robertson, M. A., Staden, R., Tanake, Y. and Brownlee, G. G. (1978). Nucleotide sequence homology at 12 intron–exon junctions in the chick ovalbumin gene. *Nature, Lond.* **275**, 510–513.
Chambon, P., Benoist, C., Breathnach, R., Cochet, M., Gannon, F., Gerlinger, P., Knist, A., Le Meur, M., Le Pennec, J. P., Mandel, J. L., O'Hare, K. and Perrin, F. (1979). Structure organization and expression of ovalbumin and related chicken genes. *In* "From Gene to Protein: Information transfer in normal and abnormal cells", Miami Winter Symposia (Eds R. Russell, K. Brew., H. Faber and J. Schultz) Vol. 16, 55–81. Academic Press, New York.
Chang, A., Nurnberg, J., Kaufman, R., Ehrlich, H., Schimke, R. and Cohen, S. (1978). Phenotypic expression in *Escherichia coli* of a DNA sequence coding for mouse dihydrofolate reductase. *Nature, Lond.* **275**, 617–624.
Chow, L. T., Gelinas, R. E., Broker, T. R. and Roberts, R. J. (1977). An amazing arrangement at the 5′ ends of adenovirus 2 messenger RNA. *Cell* **12**, 1–8.
Clarke, L. and Carbon, J. (1975). Biochemical construction and selection of hybrid plasmids containing specific segments of the *Escherichia coli* genome. *Proc. Natn. Acad. Sci. U.S.A.* **72**, 4361–4365.
Clarke, L. and Carbon, J. (1976). A colony bank containing synthetic ColEl hybrid plasmids representative of the entire *E. coli* genome. *Cell* **9**, 91–99.
Clarke, L. and Carbon, J. (1978). Functional expression of cloned yeast DNA in *Escherichia coli.* Specific complementation of argininosuccinate ligase (arg H) mutations. *J. Mol. Biol.* **120**, 517–532.
Clewell, D. B. (1974). Nature of Col El plasmid replication in *E. coli* in the presence of chloramphenicol. *J. Bacteriol.* **110**, 667–676.
Cohen, S. N., Chang, A. C. Y., Boyer, H. W. and Helling, R. B. (1973). Construction of biologically functional bacterial plasmids *in vitro. Proc. Natn. Acad. Sci. U.S.A.* **70**, 3240–3244.
Collins, J. and Brüning, H. J. (1978). Plasmids usable as gene-cloning vectors in an *in vitro* packaging by coliphage λ 'cosmids'. *Gene* **4**, 85–107.

Collins, J. and Hohn, B. (1979). Cosmids: a type of plasmid gene-cloning vector that is packageable *in vitro* in bacteriophage λ heads. *Proc. Natn. Acad. Sci. U.S.A.* **75**, 4242–4246.

Covey, C., Richardson, D. and Carbon, J. (1976). A method for the deletion of restriction enzyme sites in bacterial plasmid DNA. *Mol. Gen. Genet.* **145**, 155–158.

Craik, C. S., Buchman, S. R. and Beycho, K. (1980). Characterization of globin domains: Heme binding to the central exon product. *Proc. Natn. Acad. Sci. U.S.A.* **77**, 1384–1388.

Curtis, P. J. and Weissmann, C. (1976). Purification of globin messenger RNA from DMSO induced Friend cells and detection of a putative globin mRNA precursor. *J. Mol. Biol.* **105**, 1061–1075.

Curtiss III, R. (1976). Genetic manipulation of microorganisms. Potential benefits and biohazards. *In* "Annual Review of Microbiology" (Ed. M. P. Starr) Vol. 30, 507–530. Annual Reviews, Palo Alto, Ca.

De Wet, J. R., Daniels, D. L., Schroeder, J. L., Williams, B. G., Denniston-Thompson, K., Moore, D. D. and Blattner, F. R. (1980). Restriction maps for twenty-one charon vector phages. *J. Virol.* **33**, 401–410.

Dickson, R. (1980). Expression of a foreign eukaryotic gene in *Saccharomyces cerevisiae*: β-galactosidase from *Kluyveromyces lactis. Gene* **10**, 347–356.

Doel, M. T., Houghton, M., Cook, E. A. and Carey, N. H. (1977). The presence of ovalbumin mRNA coding sequences in multiple restriction fragments of chicken DNA. *Nucl. Acids Res.* **4**, 3701–3703.

Dugaiczyk, A., Boyer, H. W. and Goodman, H. M. (1975). Ligation of Eco RI endonuclease-generated DNA fragments into linear and circular structures. *J. Mol. Biol.* **96**, 171–184.

Dunn, A. R. and Hassell, J. A. (1977). A novel method to map transcripts: Evidence for homology between an adenovirus mRNA and discrete multiple regions of the viral genome. *Cell* **12**, 23–36.

Early, P. W., Davis, M. M., Kaback, D. B., Davidson, N. and Hood, L. (1979). Immunoglobulin heavy chain gene organization in mice: Analysis of a myeloma genomic clone containing variable and α-constant regions. *Proc. Natn. Acad. Sci. U.S.A.* **76**, 857–861.

Earnshaw, W. C. and Casjens, S. R. (1980). DNA packaging by the double-stranded DNA bacteriophages. *Cell* **21**, 319–331.

Edgell, M. H., Weaver, S., Haigwood, N. and Hutchinson III, C. A. (1979). *In* "Genetic Engineering" (Eds J. K. Setlow and A. Hollaender) Vol. 1, Plenum Press, New York and London.

Efstratiadis, A., Posakony, J. W., Maniatis, T., Lawn, R. M., O'Connell, C., Spritz, R. A., DeRiel, J. K., Forget, B., Weissman, S. M., Slightom, J. L., Blechl, A. E., Smithies, O., Baralle, F. E., Shoulders, C. C. and Proudfoot, N. J. (1980). The Structure and Evolution of the Human β-globin gene family. *Cell* **21**, 653–668.

Ehrlich, M., Cohen, S. and McDevitt, H. (1978). A sensitive radioimmunoassay for detecting products translated from cloned DNA fragments. *Cell* **13**, 681–689.

Ehrlich, S. D. (1977). Replication and expression of plasmids from *Staphylococcus aureus* in *Bacillus subtilis. Proc. Natn. Acad. Sci. U.S.A.* **74**, 1680–1682.

Feiss, M., Fisher, R. A., Crayton, M. A. and Egner, C. (1977). Packaging of bacteriophage λ chromosome: Effect of chromosome length. *Virology* **77**, 281–293.

Flavell, R. A., Kooter, J. M., de Boer, E., Little, P. F. R. and Williamson, R.

(1978). Analysis of the β—δ globin gene loci in normal human DNA: Direct determination of gene linkage and intergene distance. *Cell* **15**, 25—41.

Flavell, R. A., Grosveld, G. C., Grosveld, F. G., Bernards, R., Kooter, J. M. and De Boer, E. (1979). *In* "From gene to Protein", Miami Winter Symposium. (Eds T. R. Russell, K. Brew, H. Faber and J. Schultz) Vol. 16, 149—165, Academic Press, New York.

Fritsch, E. F., Lawn, R. M. and Maniatis, T. (1980). Molecular cloning and characterization of the human β-like globin gene cluster. *Cell* **19**, 959—972.

Gannon, F., O'Hare, K., Perrin, F., LePennec, J. P., Benoist, C. Cochet, M., Breathnach, R., Royal, A., Garapin, A., Cami, B. and Chambon, P. (1979). Organization and sequence at the 5′ end of a cloned complete ovalbumin gene. *Nature, Lond.* **278**, 428—434.

Gilbert, W. (1978). Why Genes in Pieces? *Nature, Lond.* **271**, 501.

Glover, D. M., White, R. L., Finnegan, D. J. and Hogness, D. S. (1975). Characterization of six cloned DNAs from *Drosophila melanogaster*, including one that contains the genes for rRNA. *Cell* **5**, 149—157.

Goossens, M., Dozy, A. M., Embury, S. H., Zachariades, Z., Jadjiminas, M. G., Stamatoyannopoulos, G. and Kan, Y. W. (1980). Triplicated α-globin loci in humans. *Proc. Natn. Acad. Sci. U.S.A.* **77**, 518—521.

Graham, F. and Van Der Eb, L. (1973). A new technique for the assay of infectivity of human adenovirus 5 DNA. *Virology* **52**, 1156—1167.

Grosschedl, R. and Birnstiel, M. L. (1980). Identification of regulatory sequences in the prelude sequences of an H_2 A histone gene by the study of specific deletion mutants *in vivo*. *Proc. Natn. Acad. Sci. U.S.A.* **77**, 1432—1436.

Grosveld, G. C., Koster, A. and Flavell, R. A. (1981). A transcription map of the rabbit β-globin gene. *Cell* **23**, 573—584.

Grunstein, M. and Hogness, D. (1975). Colony hybridization: A method for the isolation of cloned DNAs that contain a specific gene. *Proc. Natn. Acad. Sci. U.S.A.* **72**, 3961—3965.

Gryczan, T., Contente, S. and Dubnau, D. (1978). Characterization of *Staphylococcus aurens* plasmids introduced by transformation into *Bacillus subtilis*. *J. Bacteriol.* **134**, 218—329.

Gumport, R. I. and Lehman, I. R. (1971). Structure of the DNA ligase adenylate intermediate: lysine (ϵ-amino) linked AMP. *Proc. Natn. Acad. Sci. U.S.A.* **68**, 2559—2563.

Hanahan, D. and Meselson, M. (1980). Plasmid screening at high density. *Gene* **10**, 63—67.

Hardies, S. C. and Wells, R. D. (1976). Preparative fractionation of DNA restriction fragments by reversed phase column chromatography. *Proc. Natn. Acad. Sci. U.S.A.* **73**, 3117—3121.

Hastie, N. and Held, W. (1978). Analysis of mRNA populations by cDNA-mRNA hybrid-mediated inhibition of cell-free protein synthesis. *Proc. Natn. Acad. Sci. U.S.A.* **75**, 1217—1221.

Herrmann, R., Neugebauer, K., Pirkl, E., Zentraf, H. and Schaller, H. (1980). Conversion of bacteriophage fd into an efficient single-stranded DNA vector system. *Mol. Gen. Genet.* **177**, 231—242.

Hershey, A. D. (1971). "The Bacteriophage Lambda". Cold Spring Harbor Laboratory, New York.

Hershfield, V., Boyer, H. W., Yanofsky, C., Lovatt, N. and Helinski, D. R. (1974). Plasmid Col E1 as a molecular vehicle for cloning and amplification of DNA. *Proc. Natn. Acad. Sci. U.S.A.* **71**, 3455—3459.

Hinnen, A., Hicks, J. and Fink, G. (1978). Transformation of yeast. *Proc. Natn. Acad. Sci. U.S.A.* **75**, 1929–1933.

Hoeijmakers, J., Borst, P., Van Den Berg, J., Weissmann, C. and Cross, G. (1980). The isolation of plasmids containing DNA complementary to messenger RNA for variant surface glycoproteins of *Trypanosoma brucei. Gene* **8**, 391–417.

Hofstetter, M., Schamböck, A., Van Den Berg, J. and Weissmann, C. (1976). Specific excision of the inserted DNA segment from hybrid plasmids constructed by the poly(dA):poly(dT) method. *Biochim. Biophys. Acta.* **454**, 587–591.

Hohn, B. (1975). DNA as a substrate for packaging in bacteriophage lambda. *J. Mol. Biol.* **98**, 93–106.

Hohn, B. and Collins, J. (1980). A small cosmid for efficient cloning of large DNA fragments. *Gene* **11**, 291–298.

Hohn, B. and Hinnen, A. (1980). *In* "Genetic Engineering, Principles and Methods". (Eds J. K. Setlow and A. Hollaender) Vol. 2, 169–183. Plenum Press, New York.

Hohn, B. and Murray, K. (1978). Packaging recombinant DNA molecules into bacteriophage particles *in vitro. Proc. Natn. Acad. Sci. U.S.A.* **74**, 3259–3263.

Inoue, M. and Curtiss III, R. (1977). Transformation procedure to *E. coli* X 1776 strain. *In* "Molecular Cloning of Recombinant DNA" (Eds W. Scott and R. Werner) 248–271. Academic Press, New York.

Jackson, D. A., Symons, R. H. and Berg, P. (1972). Biochemical method for inserting new genetic information into DNA from Simian virus 40: circular SV40 DNA molecules containing lambda phage genes and the galactosidase operon of *E. coli. Proc. Natn. Acad. Sci. U.S.A.* **69**, 2904–2909.

Jeffreys, A. J. and Flavell, R. A. (1977). The rabbit β-globin gene contains a large insert in the coding sequence. *Cell* **12**, 1097–1108.

Jeffreys, A. J., Wilson, V., Wood, D., Simons, J. P., Kay, R. M. and Williams, J. G. (1980). Linkage of adult α- and β-globin genes in *X. laevis* and gene-duplication by tetraploidization. *Cell* **21**, 555–564.

Jelinek, W. R., Toomey, T. R., Leinwand, L., Duncan, C. H., Biro, P. A., Choudary, P. N., Weissman, S. M., Rubin, C. M., Houck, C. M., Deininger, P. L., and Schmid, C. W. (1980). Ubiquitous, interspersed repeated sequences in mammalian genomes. *Proc. Natn. Acad. Sci. U.S.A.* **77**, 1398–1402.

Jimenez, A. and Davies, J. (1980). Expression of a transposable antibiotic resistance element in *Saccharomyces. Nature, Lond.* **287**, 859–871.

Kahn, M., Kolter, R., Thomas, C., Figurski, D., Meyer, R., Remant, E. and Helinski, D. R. (1979). *In* "Methods in Enzymology" (Ed. R. Wu) Vol. 68, 268–280. Academic Press, New York.

Karn, J., Brenner, S., Barnett, L. and Cesareni, G. (1980). Novel bacteriophage λ cloning vector. *Proc. Natn. Acad. Sci. U.S.A.* **77**, 5172–5176.

Kedes, L. H., Chang, A. C. Y., Housman, D. and Cohen, S. N. (1975). Isolation of histone genes from unfractionated sea urchin DNA by subculture cloning in *E. coli. Nature, Lond.* **255**, 523–538.

Keggins, K., Lovett, P. and Duvall, E. (1978). Molecular cloning of genetically active fragments of *Bacillus* DNA in *Bacillus subtilis* and properties of the vector plasmid pUB 110. *Proc. Natn. Acad. Sci. U.S.A.* **75**, 1423–1427.

Kinniburgh, A. and Ross, J. (1979). Processing of the mouse β-globin mRNA precursor: At least two cleavage-ligation reactions are necessary to excise the large intervening sequences. *Cell* **17**, 915–921.

Konkel, D. A., Tilghman, S. M. and Leder, P. (1978). The evolution and sequence comparison of two recently diverged mouse chromosomal β-globin genes. *Cell* 18, 865–873.

Kressmann, A., Clarkson, S. G., Telford, J. L., and Birnstiel, M. L. (1977). Transcription of *Xenopus* tDNA^{met} and sea urchin histone DNA injected into the *Xenopus* oocyte nucleus. *In* "Cold Spring Harbor Symposia on Quantitative Biology", Vol. 42, 1077–1082. Cold Spring Harbor Laboratory, New York.

Kushner, S. (1978). "Genetic Engineering". (Eds H. W. Boyer and S. Nicosia), 17–28. Elsevier/North-Holland, Amsterdam.

Lacy, E., Hardison, R. C., Quon, D. and Mariatis, T. (1979). The linkage arrangement of four rabbit β-like globin genes. *Cell* 18, 1273–1283.

Lai, E. C., Woo, S. L. C., Bordelon-Riser, M. E., Fraser, T. H. and O'Malley, B. (1980). Ovalbumin is synthesized in mouse cells transformed with the natural chicken ovalbumin gene. *Proc. Natn. Acad. Sci. U.S.A.* **77**, 244–248.

Lauer, J., Shen, C.-K. J. and Maniatis, T. (1980). The chromosomal arrangement of human α-like globin genes: Sequence homology and α-globin gene deletions. *Cell* **20**, 119–130.

Lawn, R. M., Fritsch, E. F., Parker, R. C., Blake, G. and Maniatis, T. (1978). The isolation and characterization of linked δ- and β-globin genes from a cloned library of human DNA. *Cell* **15**, 1157–1174.

Leder, A., Miller, H. I., Hamer, D. H., Seidman, J. G., Norman, B., Sullivan, M. and Leder, P. (1978). Comparison of cloned mouse α- and β-globin genes: Conservation of intervening sequence locations and extragenic homology. *Proc. Natn. Acad. Sci. U.S.A.* **75**, 6187–6192.

Lerner, M. R., Boyle, J. A., Mount, S. M., Wolin, S. L. and Steitz, J. (1980). Are snRNPs involved in splicing? *Nature, Lond.* 283, 220–244.

Lewin, B. (1980). Alternatives for splicing. Recognising the ends of introns. *Cell* **22**, 324–326.

Lindahl, G., Sironi, G., Bialy, H. and Calender, R. (1970). Bacteriophage lambda; abortive infection of bacteria lysogenic for phage P2. *Proc. Natn. Acad. Sci. U.S.A.* **66**, 587–594.

Löfdahl, J., Sjöström, J. E. and Philipson, L. (1978). Characterization of small plasmids from *Staphylococcus aureus.* A vector for recombinant DNA in *Staphylococcus aureus. Gene* **3**, 149–159.

Lomedico, P., Rosenthal, N., Efstratiadis, A., Gilbert, W., Kolodner, R. and Tizard, R. (1979). The structure and evolution of the two non allelic rat preproinsulin genes. *Cell* 18, 545–558.

Mandel, M. and Higa, A. (1970). Calcium dependent bacteriophage DNA infection. *J. Mol. Biol.* **53**, 159–162.

Maniatis, T., Hardison, R. C., Lacy, E., Lauer, J., O'Connell, C., Quon, D., Sim, G. K. and Efstratiadis, A. (1978). The isolation of structural genes from libraries of eukaryotic DNA. *Cell* **15**, 687–701.

Maniatis, T., Fritsch, E. F., Lauer, J., and Lawn, R. M. (1980). The molecular genetics of human haemoglobins. *In* "Annual Review of Genetics", Vol. 14, 145–178. Annual Reviews, Palo Alto, Ca.

Manley, J. L., Fire, A., Cano, A., Sharp, P. A. and Gefter, M. L. (1980). DNA-dependent transcription of adenovirus genes in a soluble whole-cell extract. *Proc. Natn. Acad. Sci. U.S.A.* **77**, 3855–3859.

Mantei, N., Boll, W. and Weissmann, C. (1979). Rabbit β-globin mRNA production in mouse cells transformed with cloned rabbit β-globin chromosomal DNA. *Nature, Lond.* 281, 40–46.

Maxam, A. M. and Gilbert, W. (1980). *In* "Methods in Enzymology" (Eds

L. Grossman and K. Moldare) Vol. 65, 499–560. Academic Press, New York.

Meyerowitz, E. M., Guild, G. M., Prestidge, L. S. and Hogness, D. S. (1980). A new high-capacity cosmid and its use. *Gene* **11**, 271–282.

Mertz, J. E. and Gurdon, J. B. (1977). Purified DNAs are transcribed after microinjection into *Xenopus* oocytes. *Proc. Natn. Acad. Sci. U.S.A.* **74**, 1502–1506.

Meynell, G. G. (1972). *In* "Bacterial Plasmids" (Ed. G. G. Meynell). Macmillan, London.

Montgomery, D., Hall, B., Gillam, S. and Smith, M. (1978). Identification and isolation of the yeast cytochrome c gene. *Cell* **14**, 673–680.

Morrow, J. F. (1979). *In* "Methods in Enzymology" (Ed. R. Wu) Vol. 68, 3–26. Academic Press, New York.

Mulligan, R. and Berg, P. (1980). Expression of a bacterial gene in mammalian cells. *Science* **209**, 1422–1427.

Mulligan, R., Howard, B. and Berg, P. (1979). Synthesis of rabbit β-globin in cultured monkey kidney cells following infection with SV40 β-globin recombinant genome. *Nature, Lond.* **277**, 108–114.

Murray, N. E., Brammar, W. J. and Murray, K. (1977). Lambdoid phages that simplify recovery of *in vitro* recombinants. *Mol. Gen. Genet.* **150**, 53–61.

Nagata, S., Taira, H., Hall, A., Johnsrud, L., Streuli, M., Escödi, J., Boll, W., Cantell, K. and Weissmann, C. (1980). Synthesis in *Escherichia coli* of a polypeptide with human leukocyte interferon activity. *Nature, Lond.* **284**, 316–320.

Nishioka, Y., Leder, A. and Leder, P. (1980). Unusual α-globin-like gene that has cleanly lost both globin intervening sequences. *Proc. Natn. Acad. Sci. U.S.A.* **77**, 2806–2809.

Noyes, B., Merarech, M., Stein, R. and Aqarwal, K. (1979). Detection and partial sequence analysis of gastrin mRNA by using an oligodeoxynucleotide probe. *Proc. Natn. Acad. Sci. U.S.A.* **76**, 1770–1774.

Olby, R. (1974). *In* "The Path to the Double Helix" (Ed. R. Olby). University of Washington Press, Seattle.

Paterson, B., Roberts, B. and Kuff, E. (1977). Structural gene identification and mapping by DNA-mRNA hybrid-arrested cell-free translation. *Proc. Natn. Acad. Sci. U.S.A.* **74**, 4370–4374.

Pellicer, A., Wigler, M. Axel, R. and Silverstein, S. (1978). The transfer and stable integration of the HSV thymidine kinase gene into mouse cells. *Cell* **14**, 133–141.

Perler, F., Efstratiadis, A., Lomedico, P., Gilbert, W., Kolodner, R. and Dogson, J. (1980). The evolution of genes: the chicken preproinsulin gene. *Cell* **20**, 555–566.

Perucho, M., Hanahan, D., Lipsich, L. and Wigler, M. (1980). Isolation of the chicken thymidine kinase gene by plasmid rescue. *Nature, Lond.* **285**, 207–211.

Proudfoot, N. J. and Maniatis, T. (1980). The structure of a human α-globin pseudogene and its relationship to α-globin gene duplication. *Cell* **21**, 537–544.

Ratzkin, B. and Carbon, S. (1977). Functional expression of cloned yeast DNA in *Escherichia coli. Proc. Natn. Acad. Sci. U.S.A.* **74**, 487–491.

Reddy, V. B., Ghosh, P. K., Leibowitz, P., Piatak, M. and Weissman, S. M. (1979). Simian virus 40 early mRNAs. I. Genomic localization of 3′ and 5′ termini and two major splices in mRNA from transformed and lytically infected cells. *J. Virol.* **30**, 279–296.

Rodriguez, R. L., Bolivar, F., Goodman, H. M., Boyer, H. W. and Betlach, M. (1976). *In* "Molecular Mechanisms in the Control of Gene Expression" ICN-UCLA Symp. Mol. Cell Biol. (Eds D. P. Nierlich, W. J. Rutter and C. F. Fox) Vol. 7, 471–477. Academic Press, New York.

Roop. D. R., Tsai, M. J. and O'Malley, B. W. (1980). Definition of the 5′ and 3′ ends of transcripts of the ovalbumin gene. *Cell* **19**, 63–68.

Rothstein, R. J., Lau, L. F., Bahl, C. P., Narang, S. A. and Wu, R. (1979). *In* "Methods of Enzymology" (Ed. R. Wu) Vol. 68, 98–109. Academic Press, New York.

Rougeon, F. and Mach, B. (1977). Cloning and amplification of rabbit α- and β-globin gene sequences into *E. coli* plasmids. *J. Biol. Chem.* **252**, 2209–2217.

Sakano, H., Rogers, J. H., Huppi, K., Brack, C., Traunecker, A., Maki, R., Wall, R. and Tonegawa, S. (1979). Domains and the hinge region of an immunoglobulin heavy chain are encoded in separate DNA segments. *Nature, Lond.* **277**, 627–633.

Sanger, F., Nicklen, S. and Coulson, A. R. (1977). DNA sequencing with chain terminating inhibitors. *Proc. Natn. Acad. Sci. U.S.A.* **74**, 5463–5467.

Sanzey, B., Mercerean, O., Ternynck, T. and Kourilsky, P. (1976). Methods for identification of recombinants of phage λ. *Proc. Natn. Acad. Sci. U.S.A.* **73**, 3394–3397.

Sargent, T. D., Wu, J. R., Sala-Trepat, J. M., Wallace, R. B., Reyes, A. A. and Bonner, J. (1979). The rat serum albumin gene: analysis of cloned sequences. *Proc. Natn. Acad. Sci. U.S.A.* **76**, 3256–3260.

Schibler, U., Tosi, M., Pittet, A. C., Fabiai, L. and Wellauer, P. K. (1980). Tissue-specific expression of mouse α-amylase genes. *J. Mol. Biol.* **142**, 93–116.

Schroder, J. L. and Blattner, F. (1978). Least-squares method for restriction mapping. *Gene* **4**, 167–174.

Sgaramella, V., Van De Sande, J. H. and Khorana, M. G. (1970). Studies on polynucleotides, C. A novel joining reaction catalysed by the T4-polynucleotide ligase. *Proc. Natn. Acad. Sci. U.S.A.* **67**, 1468–1475.

Shalka, A. and Shapiro, L. (1976). *In situ* immunoassays for gene translation products in phage plaques and bacterial colonies. *Gene* **1**, 65–79.

Shen, C.-K. J. and Maniatis, T. (1980). The organization of repetitive sequences in a cluster of rabbit β-like globin genes. *Cell* **19**, 379–392.

Shortle, O. and Nathans, D. (1978). Local mutagenesis: A method for generating viral mutants with base substitutions in preselected regions of the viral genome. *Proc. Natn. Acad. Sci. U.S.A.* **75**, 2170–2174.

Sinsheimer, R. L. (1977). Recombinant DNA. *In* "Annual Review of Biochemistry" Vol. 46, 415–438. Annual Reviews, Palo Alto, Ca.

Slightom, J. L., Blechl, A. E. and Smithies, O. (1980). Human Fetal $^{G}\gamma$- and $^{A}\gamma$-globin genes: Complete nucleotide sequences suggest that DNA can be exchanged between these duplicated genes. *Cell* **21**, 627–638.

Smith, D. F., Searle, P. F. and Williams, J. G. (1979). Characterization of bacterial clones containing DNA sequences derived from *Xenopus laevis. Nucl. Acids Res.* **6**, 487–506.

Sobel, M., Yamamoto, T., Adams, S., DiLanro, R., Avvedimento, V., De Crombrugghe, B. and Pastan, I. (1978). Construction of a recombinant bacterial plasmid containing a chick pro-α2 collagen gene sequence. *Proc. Natn. Acad. Sci. U.S.A.* **75**, 5846–5850.

Soberon, X., Covarrubias, L. and Bolivar, F. (1980). Construction and characterization of new cloning vehicles. IV. Deletion derivatives of pBR322 and pBR325. *Gene* **9**, 287–305.

Southern, E. M. (1975). Detection of specific sequences among DNA fragments separated by gel electrophoresis. *J. Mol. Biol.* **98**, 503–517.

Spofford, J. (1976). *In* "The Genetics and Biology of *Drosophila*" (Eds M. Ashburner and E. Novitski) Vol. 1, 955–1018. Academic Press, London and New York.

Sternberg, N., Tiemeier, D. and Enquist, L. (1977). *In vitro* packaging of a λ *Bam* vector containing Eco RI DNA fragments of *Escherichia coli* and phage P1. *Gene* **1**, 255–280.

St John, T. and Davis, R. (1979). Isolation of galactose-inducible DNA sequences from *Saccharomyces cerevisiae* by differential plaque filter hybridization. *Cell* **16**, 443–452.

Struhl, K., Cameron, J. R. and Davis, R. W. (1976). Functional genetic expression of eukaryotic DNA in *E. coli. Proc. Natn. Acad. Sci. U.S.A.* **73**, 1471–1475.

Sugino, A., Goodman, H. M., Heyneker, H. L., Shine, J., Boyer, H. W. and Cozzarelli, N. R. (1977). Interaction of bacteriophage T4 RNA and DNA ligase in joining of duplex DNA at base-paired ends. *J. Biol. Chem.* **252**, 3987–3995.

Taniguchi, T., Sakoi, M., Fujii-Kuriyama, Y., Muramatsu, M., Kobayashi, S. and Sudo, T. (1979). Construction and identification of a bacterial plasmid containing the human fibroblast interferon gene sequence. *Proc. Jap. Acad. Sci. Ser. B* **55**, 464–469.

Tiemeier, D., Enquist, L. and Leder, P. (1976). Improved derivative of a phage λ EK2 vector for cloning recombinant DNA. *Nature, Lond.* **263**, 526–529.

Tilghman, S. M. Tiemeier, D. C., Polsky, F., Edgell, M. H., Seidman, J. G., Leder, A., Enquist, L. W., Norman, B. and Leder, P. (1977). Cloning specific segments of the mammalian genome: Bacteriophage λ containing mouse globin and surrounding gene sequences. *Proc. Natn. Acad. Sci. U.S.A.* **74**, 4406–4410.

Tilghman, S. M., Curtis, P. J., Tiemeier, D. C., Leder, P. and Weissmann, C. (1978a). The intervening sequence of a mouse β-globin gene is transcribed within the 15S globin mRNA precursor. *Proc. Natn. Acad. Sci. U.S.A.* **75**, 1309–1313.

Tilghman, S. M., Tiemeier, D. C., Seidman, J. G., Peterlin, B. M., Sullivan, M., Maizel, J. V. and Leder, P. (1978b). Intervening sequence of DNA identified in the structural portion of the mouse β-globin gene. *Proc. Natn. Acad. Sci. U.S.A.* **75**, 725–729.

Timmis, K., Cabello, F. and Cohen, S. N. (1974). Utilization of two distinct modes of replication by a hybrid plasmid constructed *in vitro* from separate replicons. *Proc. Natn. Acad. Sci. U.S.A.* **71**, 4556–4560.

Twigg, A. J. and Sherratt, D. (1980). Trans-complementable copy-number mutants of plasmid Col EI. *Nature, Lond.* **283**, 216–218.

Ullrich, A., Shine, J., Chirgwin, J., Pictet, R., Tisher, E., Rutter, W. J. and Goodman, H. M. (1977). Rat insulin genes: construction of plasmids containing the coding sequences. *Science* **196**, 1313–1316.

Upcroft, D., Sholnik, H. Upcroft, J., Solomon, D., Khoury, G., Hamer, D. and Fareed, G. (1978). Transduction of a bacterial gene into mammalian cells. *Proc. Natn. Acad. Sci. U.S.A.* **75**, 2117–2121.

Van Den Berg, J., Van Ooyen, A., Mantei, N., Schamböck, A., Grosveld, G. C., Flavell, R. A. and Weissmann, C. (1978). Comparison of cloned rabbit and mouse β-globin genes showing strong evolutionary divergence of two homologous pairs of introns. *Nature, Lond.* **275**, 37–44.

Van Der Ploeg, L. H. T., Groffen, J. and Flavell, R. A. (1980). A novel type of secondary modification of two CCGG residues in the human $\gamma\delta\beta$-globin gene locus. *Nucl. Acids Res.* **8**, 4563–4574.

Vapnek, D., Hantala, J., Jacobson, J., Giles, N. and Kushner, S. (1977). Expression in *Escherichia coli* K12 of the structural gene for catabolic dehydroginase of *Neurospora crassa*. *Proc. Natn. Acad. Sci. U.S.A.* **74**, 3508–3512.

Villa-Komaroff, L., Efstradiatis, A., Broome, S., Lomedico, P., Tizard, R., Naber, S. P., Chick, W. L. and Gilbert, W. (1978). A bacterial clone synthesizing proinsulin. *Proc. Natn. Acad. Sci. U.S.A.* **75**, 3727–3731.

Villareal, L. and Berg, P. (1977). Hybridization *in situ* of SV40 plaques. Detection of recombinant SV40 virus carrying specific sequences of non-viral DNA. *Science* **196**, 183–185.

Vogt, V. M. (1973). Purification and further properties of a single-strand-specific nuclease from *Aspergillus oryzae*. *Eur. J. Biochem.* **33**, 192–200.

Wahli, W., Dawid, I. B., Wyler, T., Weber, R. and Ryffel, G. U. (1980). Comparative analysis of the structural organization of two closely related vitellogenin genes in *X. laevis*. *Cell* **20**, 107–117.

Weaver, R. F. and Weissmann, C. (1979). Mapping of RNA by a modification of the Berk-Sharp procedure: The 5′ termini of 15S β-globin mRNA precursor and mature 10S β-globin mRNA have identical map coordinates. *Nucl. Acids Res.* **6**, 1175–1193.

Weil, P. A., Luse, D. S., Segall, J. and Roeder, R. G. (1979). Selective and accurate initiation of transcription at the Ad2 major late promotor in a soluble system dependent on purified RNA polymerase II and DNA. *Cell* 18, 469–484.

Weissmann, C. (1978). Reverse genetics. *Trends Biochem. Sci.* **3**, N109–N111.

Wensink, P. C., Finnegan, D. J., Donelson, J. E. and Hogness, D. S. (1974). A system for mapping DNA sequences in the chromosome of *Drosophila melanogaster*. *Cell* 3, 315–325.

White, R. L. and Hogness, D. S. (1977). R-loop mapping of the 18S and 28S sequences in the long and short repeating units of *Drosophila melanogaster* rDNA. *Cell* **10**, 177–191.

Wigler, M., Silverstein, S., Lee, L-S., Pellicer, A., Cheng, Y. and Axel, R. (1977). Transfer of purified herpes simplex virus thymidine kinase gene to cultured mouse cells. *Cell* **11**, 223–232.

Wigler, M., Pellicer, A., Silverstein, S. and Axel, R. (1978). Biochemical transfer of single-copy eukaryotic genes using total cellular DNA as donor. *Cell* **14**, 725–731.

Wigler, M., Sweet, R., Sim, G. K., Wold, B., Pellicer, A., Lacy, E., Maniatis, T., Silverstein, S. and Axel, R. (1979). Transformation of mammalian cells with genes from procaryotes and eucaryotes. *Cell* **16**, 777–785.

Williams, B. G. and Blattner, F. R. (1979). Construction and characterization of the hybrid bacteriophage lambda charon vectors for DNA cloning. *J. Virol.* **29**, 555–575.

Wold, B., Wigler, M., Lacy, E., Maniatis, T., Silverstein, S. and Axel, R. (1979). Introduction and expression of a rabbit β-globin gene in mouse fibroblasts. *Proc. Natn. Acad. Sci. U.S.A.* **76**, 5684–5688.

Woolford, J. and Rosbash, M. (1979). The use of R-looping for structural gene identification and mRNA purification. *Nucl. Acids Res.* **6**, 2483–2497.

Young, R. A., Hagenbuchle, O. and Schibler, U. (1981). A single mouse α-amylase gene specifies two different tissue specific mRNAs. *Cell*, in press.

The use of restriction enzymes in genetic engineering

A. D. B. MALCOLM

Department of Biochemistry, St Mary's Hospital Medical School, University of London, London, UK

I Introduction

A The restriction phenomenon

When the coliphage λ is grown on *E. coli* strain C and then attempts are made to grow it in *E. coli* K12, it grows very poorly (Bertani and Weigle, 1953). It was shown that this was caused by a response of the host to the phage. Since the growth of the phage was restricted, the mechanism used by the *E. coli* was called a "restriction system". Phage grown originally on K12 could be regrown efficiently on K12, and this implies that the host also possessed a means of protecting the phage from this restriction and the mechanism became a "restriction—modification system".

Meselson and Yuan (1968) incubated λ DNA, which had been grown in *E. coli* K12, and λ DNA from *E. coli* C with an extract from K12. Analysis by sedimentation through a sucrose gradient showed that the λ.K DNA was unaffected but that λ.C DNA was degraded. This degradation was limited in extent and suggested that the restriction system (enzyme) only hydrolysed the DNA at a small number of sites. As it happens the restriction enzyme from *E. coli* K does not cut the DNA at specific sequences and hence has not been greatly exploited for genetic engineering.

The first site specific restriction enzyme was discovered by Smith and Wilcox (1970) during their studies on recombination in *Haemophilus influenzae.* They showed by viscometry that a *Haemophilus* cell extract degraded DNA from the bacteriophage P22 but had no effect on DNA from *Haemophilus* itself.

Using the purified enzyme Kelly and Smith (1970) showed that the breaks produced in T7 DNA all occurred at the symmetrical sequence

$$5' \ldots\ldots \mathrm{G\,p\,T\,p\,Py}^{\downarrow}\mathrm{p\;\;Pu\,p\,A\,p\,C} \ldots\ldots 3'$$

$$3' \ldots\ldots \mathrm{C\,p\,A\,p\,Pu\;\;p}_{\uparrow}\mathrm{Py\,p\,T\,p\,G} \ldots\ldots 5'$$

where the hydrolysis occurs at the phosphodiester bonds indicated by the arrows.

A catalogue of restriction endonucleases isolated since then now numbers over 250 examples, although it is not certain that all these are in fact different enzymes (Table 1). Excellent earlier reviews are by Roberts (1976) and Modrich (1979).

II The enzymes

A Nomenclature

Strictly speaking the term "restriction endonucleases" refers to those enzymes which have been demonstrated (either biochemically or genetically) to be part of a restriction—modification system. In practice this has only been achieved for a handful of such enzymes and it is now usual to refer to any sequence specific endonuclease as a restriction enzyme.

The abbreviated name (Smith and Nathans, 1973) for each enzyme is derived from the parent organism — thus Eco from *Escherichia coli*, Sal from *Streptomyces albus* etc. If necessary the serotype or strain is then identified — Hinc and Hind from *Haemophilus influenzae* strains c and d. Finally a number is added which simply reflects chronological order of isolation — Acc I and Acc II from *Acinetobacter calcoaceticus* etc.

Restriction enzymes are also categorized as Type I, II or III.

Type I restriction enzymes do not cleave DNA at specific sequences (Arber, 1974) and for this reason have not been particularly useful in genetic engineering. They require ATP, Mg^{2+} and *S*-adenosyl methionine for activity. These enzymes can probably correctly be considered as multi-enzyme complexes since they are large (molecular weights around 300 000), composed of three non-identical subunits and have several different activities including DNA methylation, DNA cleavage and ATP hydrolysis.

Type III enzymes (Kauc and Piekarowicz, 1978) also require ATP and Mg^{2+} but do not have an absolute requirement for SAM although they are stimulated by it.

It is the Type II restriction enzymes which cleave at or very close to a defined recognition sequence which have been extensively used in genetic engineering and which will therefore be the subject of the rest of this article.

B Purification

Few restriction enzymes have been purified to the extent that single protein peaks from a chromatographic column or single bands on a polyacrylamide gel can be identified with enzyme activity. Nonetheless a large number has been purified free of contaminating nucleases, phosphatases, etc. which would interfere with their use for genetic engineering. Naturally each enzyme will require its own variations but it may be useful to outline some general principles (Pirrotta and Bickle, 1980).

Cells may be broken open by any of a variety of methods: sonications; lysozyme and deoxycholate treatment in a Waring blender; grinding with glass beads; or decompression in a French press. High speed centrifugation (e.g. 100 000 g for 60 min) will suffice to pellet cell debris. Nucleic acid may be removed from the supernate by precipitation with polyethyleneimine which, being polycationic, binds polyanions tightly (Burgess and Jendrisak, 1975; Bickle *et al.*, 1977). Careful adjustment of ionic strength may be required at this stage to prevent the enzyme binding to the DNA and co-precipitating.

Affinity chromatography on immobilized Cibacron F3G-A has proved useful at an early state (Baksi *et al.*, 1978; Baksi and Rushizky, 1979). This ligand appears to display a preference for enzymes whose substrates contain stacked nucleotides (Thompson *et al.*, 1975; Graham *et al.*, 1976). Subsequent chromatography on phosphocellulose or heparin—agarose (both of which being polyanionic preferentially bind enzymes whose substrates are polyanoinic) is often sufficient to give a usable preparation, although a final sizing step by gel filtration may be necessary.

Further details of particular enzymes can be obtained from the references given in Table 1.

C Assay

Owing to the high molecular weight of DNA, its solutions are appreciably viscous. A reduction in molecular weight caused by nuclease digestion produces a drop in viscosity and this was used for the earliest assays of restriction enzyme activity (Smith and Wilcox, 1970). This method however lacks the precision obtainable by gel electrophoresis (Sharp *et al.*, 1973). Since DNA is negatively charged it will migrate towards the anode. Its high molecular weight means that a matrix with large pores such as agarose at a concentration between 0.5% and 1.5% is appropriate, although for lower molecular weight fragments mixed agarose—polyacrylamide is suitable (Southern, 1979). The combined effect of size and charge results in the mobility being proportional to the logarithm of the molecular weight (Fig. 1) although discontinuous or gradient concentration gels may be used to provide other separations (Jeppesen, 1980).

The DNA bands are then visualized by staining the gel in a solution of ethidium bromide (CARE, ethidium bromide is thought to be carcinogenic) followed by destaining. This phenanthridine derivative (Fig. 2) binds to DNA both ionically and by intercalation. Under UV light it fluoresces strongly enabling as little as 10 ng of DNA in a band to be seen (Fig. 3). Staining with ethidium bromide prior to electrophoresis suffers from the disadvantage that intercalation causes the double helix to increase in length, thus reducing the DNA's mobility.

Alternatively if the DNA is radioactive (preferably labelled with ^{32}P) the bands may be visualized by autoradiography. The sensitivity of this method depends on the extent of labelling and on the specific activity of the ^{32}P used but 100 pg of DNA in a band should be detectable after 15 hours' autoradiography.

Table 1 All enzymes for which either the recognition sequence is known or for which a published reference work exists (see also Roberts, 1981, for other unpublished work).

Enzyme	Sequence	Reference	Commercially available
Aac I	GGATCC	—	
Acc I	GT↓$^{AG}_{CT}$AC	—	Yes
Acc II	CGCG	—	Yes
Atu BI	CC$^{A}_{T}$GG	Roizes *et al.* (1977)	
Atu BVI	—	Roizes *et al.* (1979)	
Atu II	CC$^{A}_{T}$GG	Lebon *et al.* (1978)	
Atu CI	TGATCA	—	
Acy I	GPu↓CGPyC	DeWaard *et al.* (1978)	
Aos I	TGC↓GCA	DeWaard *et al.* (1979)	
Aos II	GPu↓CGPyC	DeWaard *et al.* (1979)	
Ast WI	GPu↓CGPyC	—	
Asu I	G↓GNCC	Hughes *et al.* (1980)	
Asu II	TT↓CGAA	—	
Asu III	GPu↓CGPyC	—	
Ava I	CPyCGPuG	Murray *et al.* (1976)	Yes
Ava II	G↓G$^{A}_{T}$CC	Murray *et al.* (1976)	Yes
Ava III	ATGCAT	Roizes (1979)	
Avr I	CPyCGPuG	—	
Avr II	CCTAGG	—	Yes
Alu I	AG↓CT	Roberts *et al.* (1976)	Yes
Apy I	CC↓$^{A}_{T}$GG	—	
Bac I	CCGCGG	—	
Bam FI	GGATCC	Shibata *et al.* (1976)	
Bam HI	G↓GATCC	Wilson and Young (1980)	Yes
Bam N	G↓G$^{A}_{T}$CC	Shibata and Ando (1975)	
Bbv I	GC$^{A}_{T}$GC	Vanyushin and Dobritsa (1975)	Yes
Bcl I	T↓GATCA	Bingham *et al.* (1978)	Yes
Bcel 70	CTGCAG	Shibata *et al.* (1976)	
Bce R	CGCG	Shibata *et al.* (1976)	
Bgl I	GCCNNNN↓NGGC	Duncan *et al.* (1978)	Yes
Bgl II	A↓GATCT	Duncan *et al.* (1978)	Yes
Bpu I	—	Ikawa *et al.* (1976)	
Bsp 1286	—	Shibata *et al.* (1976)	
Bsp RI	GGCC	Kiss *et al.* (1977)	
Bst I	G↓GATCC	Catterall and Welker (1980)	
Bst EII	G↓GTNACC	—	Yes
Bst EIII	GATC	—	
Bst PI	G↓GTNACC	Pugatsch and Weber (1979)	
Bst NI	CC↓GG	—	Yes
Bse I	GGCC	—	
Bse II	GTTAAC	—	
Bsu RI	GG↓CC	Bron and Hörz (1980)	

Enzyme	Sequence	Reference	Commercially available
Bsu 1076	GGCC	Shibata *et al.* (1976)	
Bsu 1247	CTGCAG	Shibata *et al.* (1976)	
Bbr I	AAGCTT	—	
Bpe I	AAGCTT	Greenaway (1980)	
Bal I	TGG↓CCA	Gelinas *et al.* (1977)	Yes
Blu I	C↓TCGAG	Gingeras *et al.* (1978)	
Blu II	GGCC	—	
Cla I	AT↓CGAT	—	
Clm II	GG$^{A}_{T}$CC	—	
Clt I	GG↓CC	—	
Cau I	GG$^{A}_{T}$CC	—	
Cau II	CC↓$^{G}_{C}$GG	—	Yes
Cvi I	GG↓CC	—	
Cfo I	GCG↓C	Makula and Meagher (1980)	Yes
Chu I	AAGCTT	—	
Chu II	GTPyPuAC	—	
Cpe I	TGATCA	—	
Dde I	C↓TNAG	Makula and Meagher (1980)	Yes
Dde II	CTCGAG	—	
Dds I	GGATCC	—	
Dpn I	G^{m}A↓TC	Lacks (1980)	Yes
Dpn II	GATC	Lacks (1980)	
Ecl I	—	Hartmann and Goebel (1977)	
Ecl II	CC$^{A}_{T}$GG	Hartmann and Goebel (1977)	
Eca I	G↓GTNACC	Hobom *et al.* (1981)	
Ecc I	CGGCGG	—	
Eco RI	G↓AATTC	Greene *et al.* (1974)	Yes
Eco RI*	↓AATT	Polisky *et al.* (1975)	Yes
Eco RI′	PuPuA↓TPyPy	—	
Eco RII	↓CC$^{A}_{T}$GG	—	Yes
Fsp AI	G↓GTNACC	—	
Fnu AI	G↓ANTC	Lui *et al.* (1979)	
Fnu CI	GATC	—	
Fnu DI	GG↓CC	—	
Fnu DII	CG↓CG	—	Yes
Fnu DIII	GCG↓C	—	
Fnu 4HI	GC↓NGC	Leung *et al.* (1979)	Yes
Gdi I	AGG↓CCT	—	
Gdi II	Py↓GGCCG	—	
Gdo I	GGATCC	—	
Hae I	$^{A}_{T}$GG↓CC$^{A}_{T}$	—	
Hae II	PuGCGC↓Py	Roberts *et al.* (1975)	Yes
Hae III	GG↓CC	Middleton *et al.*	Yes
Hap II	C↓CGG	Takanami (1974)	
Hga I	GACGC	Takanami (1974)	Yes

Table 1 (*Continued*)

Enzyme	Sequence	Reference	Commercially available
Hhg I	GGCC	—	
Hha I	GCG↓C	Roberts *et al.* (1976)	Yes
Hha II	GANTC	Mann *et al.* (1978)	
Hin GUI	GCGC	—	
Hinb III	AAGCTT	—	
Hin 10561	CGCG	—	
Hinc II	GTPyPuAC	Landy *et al.* (1974)	
Hind II	GTPy↓PuAC	Smith and Wilcox (1970)	
Hind III	A↓AGCTT	Old *et al.* (1975)	Yes
Hinf I	G↓ANTC	—	Yes
Hinf II	AAGCTT	—	
Hin HI	PuGCGCPy	Takanami (1974)	
Hph I	GGTGA	—	Yes
Hpa I	GGT↓AAC	Sharp *et al.* (1973)	Yes
Hpa II	C↓CGG	Sharp *et al.* (1973)	Yes
Hsu I	A↓AGCTT	—	
Hgi AI	G$^{A}_{T}$GC$^{A}_{T}$↓C	Brown *et al.* (1980)	Yes
Hgi CI	G↓GPyPuCC	—	
Hgi CII	G↓G$^{A}_{T}$CC	—	
Hgi CIII	G↓TCGAC	—	
Hgi DI	GPu↓CGPyC	—	
Hgi EII	ACC(N)$_6$GGT	—	
Kpn I	GGTA↓CC	Smith *et al.* (1976)	Yes
Mla I	TT↓CGAA	Duyvesteyn and De Waard (1980)	
Mst I	TGCGCA		Yes
Mbo I	↓GATC	Gelinas (1977)	Yes
Mbo II	GAAGA	Gelinas (1977)	
Mki I	AAGCTT	—	
Mno I	C↓CGG	—	
Mno III	GATC	—	
Mnl I	CCTC↓(N)$_5$	—	Yes
Mnn I	GTPyPuAC	—	
Mnn IV	GCGC		
Mos I	GATC	Gelinas *et al.* (1977)	
Mph I	CC$^{A}_{T}$GG	—	
Msp I	C↓CGG	—	Yes
Msi I	CTCGAG	—	
Nci I	CC$^{C}_{G}$GG	Watson *et al.* (1980)	Yes
Ngo I	PuGCGCPy	—	
Ngo II	GGCC	Clanton *et al.* (1978)	
Oxa I	AGCT	—	
Pvu I	CGAT↓CG	—	Yes
Pvu II	CAG↓CTG	—	Yes
Pal I	GGCC	—	
Pst I	CTGCA↓G	Smith *et al.* (1976)	Yes
Pae R7	—	Hinkle and Milter (1979)	Yes
Pta I	GATC	—	

Table 1 *(Continued)*

Enzyme	Sequence	Reference	Commercially available
Pma I	CTGCAG	—	
Rru I	AGT↓ACT	—	
Rru II	CC↓$^{A}_{T}$GG	—	
Rsh I	CGAT↓CG	Lynn *et al.* (1979)	
Rsa I	GT↓AC	Lynn *et al.* (1980)	Yes
Rsr I	GAATTC	—	
Sma I	CCCGGG	Endow and Roberts (1977)	Yes
Sna I	GTATAC	—	
Sci NI	G↓CGC	—	
Sau 3A	↓GATC	Sussenbach *et al.* (1976)	Yes
Sau 96I	G↓GNCC	Sussenbach *et al.* (1978)	Yes
Sfa I	GG↓CC	Wu *et al.* (1978)	
Sfa GU1	CCGG	—	
Sfa NI	GATGC	—	
Sac I	GAGCT↓C	—	Yes
Sac II	CCGCGG	—	Yes
Sal PI	CTGCAG	Chater (1977)	
Spa I	CTCGAG	—	
Sal I	G↓TCGAC	Arrand *et al.* (1978)	
Sau I	CC↓TNAGG	Timko *et al.* (1981)	
Sbo I	CCGCGG	Takahashi *et al.* (1979)	
Scu I	CTCGAG	Takahashi *et al.* (1979)	
Sfr I	CCGCGG	Takahashi *et al.* (1979)	
Sgo I	CTCGAG	—	
Shy I	CCGCGG	Walter *et al.* (1978)	
Sla I	C↓TCGAG	Takahashi *et al.* (1979)	
Sph I	GCATG↓C	Fuchs *et al.* (1980)	
Sst I	GAGCT↓C	Goff and Rambach (1978)	Yes
Sst II	CCGC↓GG	Goff and Rambach (1978)	Yes
Sst IV	TGATCA	Hu *et al.* (1978)	
Stu I	AGG↓CCT	Shimotsu *et al.* (1980)	
Tha I	CG↓CG	McConnell *et al.* (1978)	Yes
Tgl I	CCGCGG	—	
Taq I	T↓CGA	Sato *et al.* (1977)	Yes
Tfl I	TCGA	Sato *et al.* (1980)	
TthHB8 I	TCGA	Sato *et al.* (1977)	
Tth III I	GAC$(N)_3$GTC	Shinomiya and Sato (1980)	
Tth III II	CAAPuCA	Shinomiya *et al.* (1980)	
Ttn I	GGCC	—	
Xam I	GTCGAC	Arrand *et al.* (1978)	
Xba I	T↓CTAGA	Zain and Roberts (1977)	Yes
Xho I	C↓TCGAG	Gingeras *et al.* (1978)	Yes
Xho II	Pu↓GATCPy	—	
Xma I	C↓CCGGG	Endow and Roberts (1977)	Yes
Xma II	CTGCAG	Endow and Roberts (1977)	
Xma III	CGGCCG	Kunkel *et al.* (1979)	Yes
Xni I	CGATCG	—	
Xor I	CTGCAG	Wang *et al.* (1980)	
Xor II	CGATC↓G	Wang *et al.* (1980)	Yes
Xpa I	C↓TCGAG	Gingeras *et al.* (1978)	

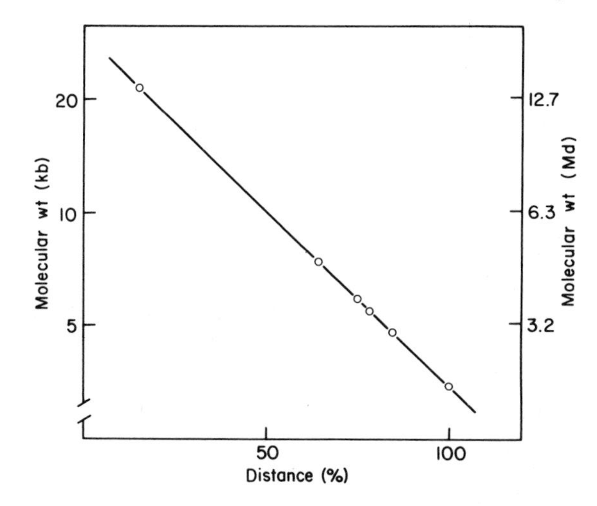

Figure 1 The distance migrated by Eco RI generated fragments of bacteriophage λ DNA during electrophoresis plotted against molecular weight (on a logarithmic scale).

Quantitation of both these methods is probably best achieved using internal standards.

D Kinetics

In most cases the genetic engineer merely wishes to achieve a total (or limit) digest of a particular DNA sample. Sometimes a partial digest is required (for example if it is desired to produce a larger piece of DNA containing the desired sequence). Usually the approach

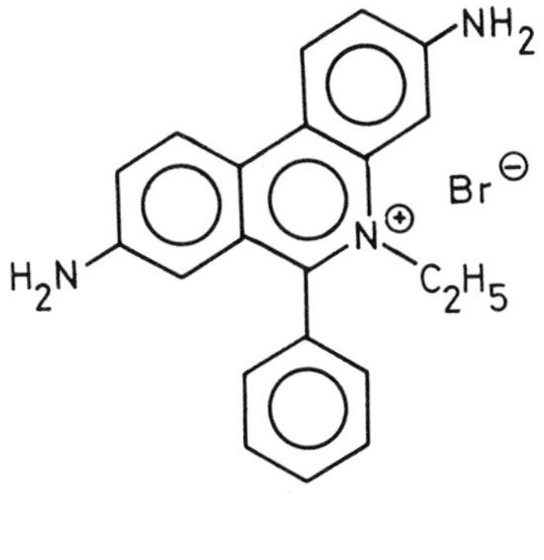

Figure 2 Ethidium bromide.

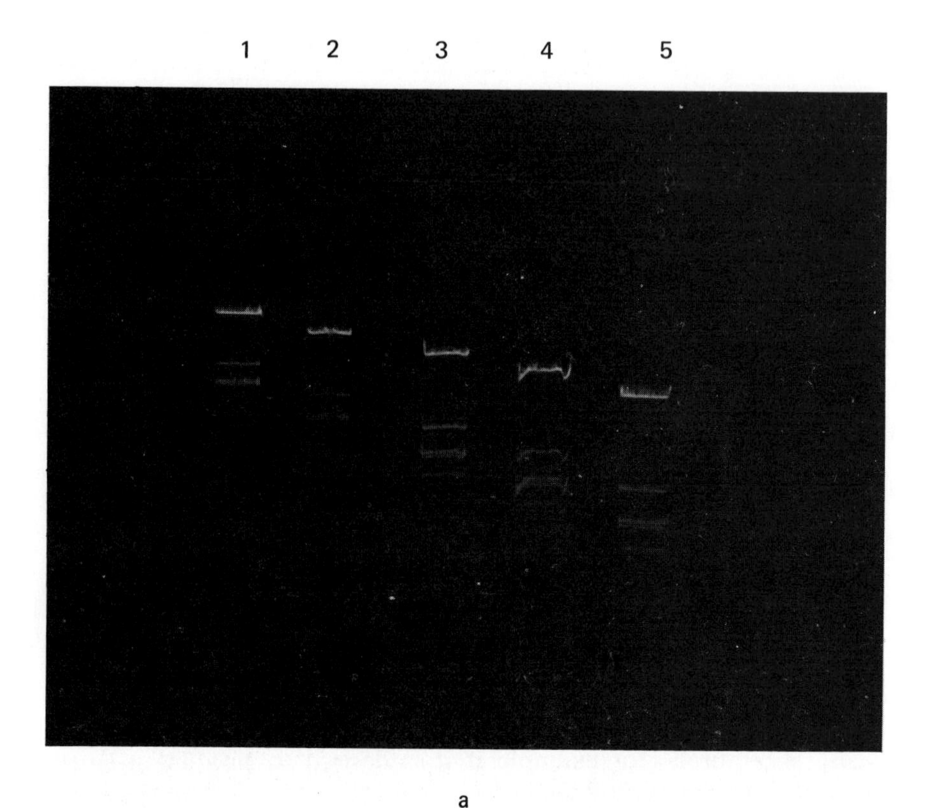

Figure 3a The separation of Eco RI generated fragments of bacteriophage DNA by electrophoresis through 0.7% agarose at 30 V for varying times. Track 1, 130 min; Track 2, 160 min; Track 3, 190 min; Track 4, 210 min; Track 5, 240 min.

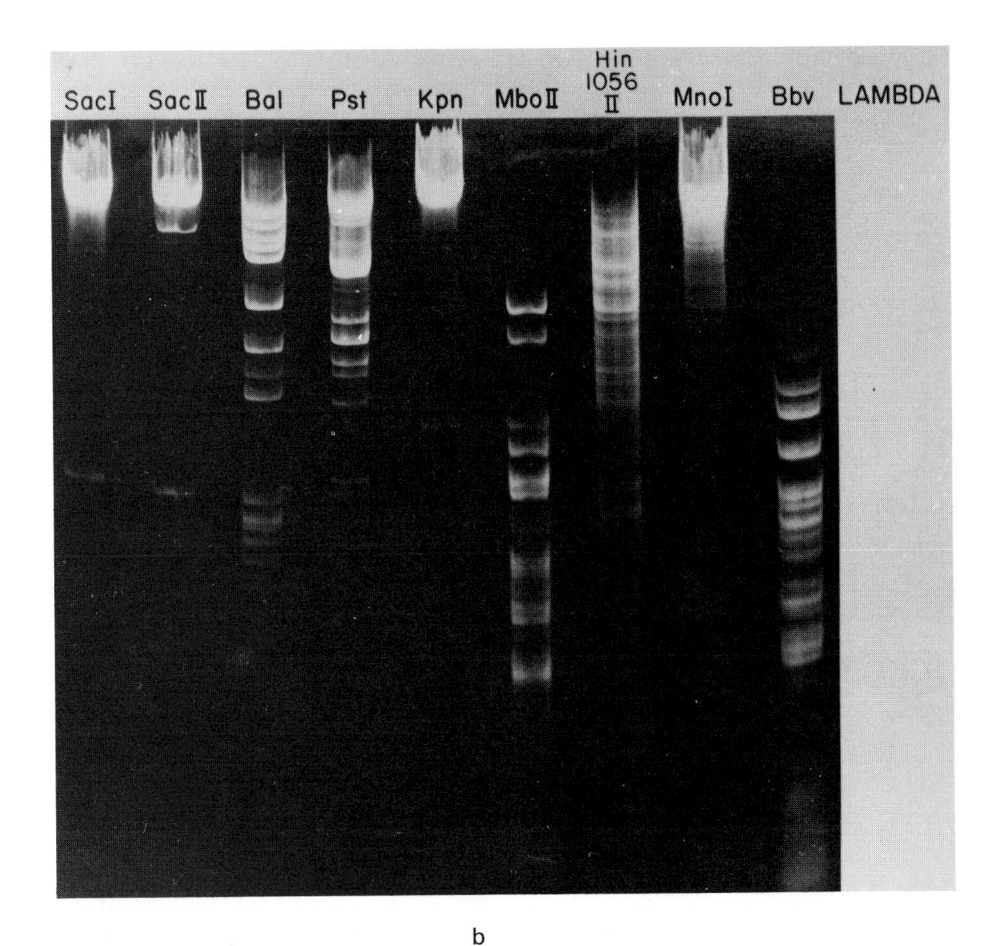

b

Figure 3b Pattern of fragments obtained by restricting phage DNA with various enzymes. The DNA was stained with ethidium bromide after electrophoresis and photographed under UV light.

to a partial digest is closer to "hit or miss" rather than carefully calculated. However it may be useful to consider some of the factors influencing the kinetics of restriction enzymes.

Although it is generally assumed that restriction enzymes behave in a simple Michaelis-Menten type manner (mainly because the lack of data means there is nothing to contradict this view), there are nevertheless several complications:

(1) The Michaelis-Menten equation is derived assuming that $[S] \gg [E]$. Because the quantities of DNA used are very small and the molecular weight per recognition site is very large, this may not always be the case. More often $[S] > [E]$ or even $[S] \approx [E]$.

(2) It will normally be the case that $[S]_0 \ll K_m$ rather than $[S]_0 \gg K_m$. This will mean that

$$v = \frac{d[P]}{dt} = \frac{k[E]_0\,[S]_0}{K_m + [S]_0}$$

becomes

$$\frac{d[P]}{dt} = \frac{V[S]_0 - [P]}{K_m + [S]_0 - [P]}$$

and if $K_m \gg [S]_0$ this may be simplified to

$$\frac{d[P]}{dt} = \frac{V([S]_0 - [P]}{K_m}$$

which on integration gives

$$t = \frac{K_m}{V} \ln \frac{[S]_0}{[S]_0 - [P]}$$

Thus the time for 50% digestion is 0.69 K_m/V and for 95% digestion is 3.0 K_m/V. Figure 4 summarizes such calculation. It is perhaps worth emphasizing that the time required for 99% digestion is double that for 90% digestion.

(3) Different sites for the same enzyme are digested at different rates. For example the rates of digestion of the five Eco RI sites in phage λ DNA vary over a ten-fold range (Thomas and Davis, 1975).

If we consider a molecule with just two sites A and B for an enzyme, the integrated rate equation is

$$t = \frac{A_0 - A}{V_a} + \frac{K_a}{V_a} \ln \frac{A_0}{A} + \frac{B_0 K_a}{V_a K_b A_0^r r} [A_0 - A]$$

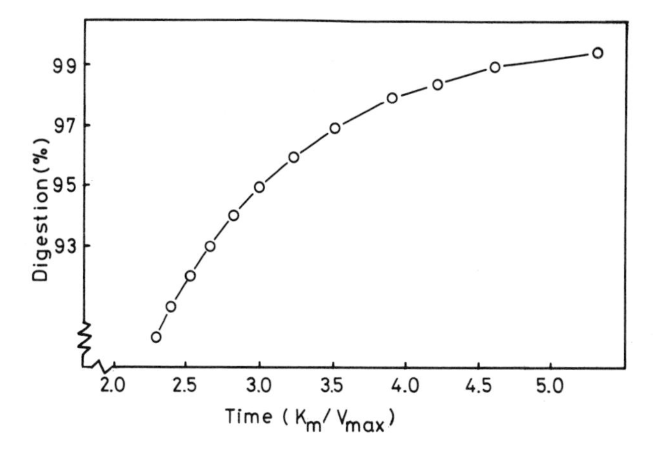

Figure 4 Predicted time course of the later part of a restriction enzyme digestion.

where

$$r = \frac{V_b K_a}{V_a K_b}$$

If r is very large ($\geqslant 10$) or very small ($\leqslant 0.1$), then the effect may be ignored. Similarly for $A \geqslant 0.9A_0$. However for values of A around $0.5A_0$ (50% digestion) it may be worthwhile to allow for this.

Restriction enzymes will bind to polydeoxynucleotides which do not contain the cleavage sequence (Woodhead and Malcolm, 1980) which results in inhibition of the enzyme (Langowski *et al.*, 1980). Since any natural DNA will always contain both types of site, any increase in substrate concentration will always result in an increase in inhibitor concentration. This phenomenon has been analysed by Langowski *et al.* (1980) and assuming that the conditions of the Michaelis-Menten equation are valid (but see observations above) they have shown that

$$v = \frac{V_a K_i}{K_i + lK_m} \cdot \frac{[S]}{K_i K_m /(K_i + lK_m) + [S]}$$

where V_m and K_m are the true maximum velocity and Michaelis constant (i.e. in the absence of the inhibitory DNA), K_i is the inhibitory constant of the non-specific DNA, l is the relative concentration of non-specific "sites" (i.e. lengths of DNA equivalent to the true recognition site) and $[S]$ is the concentration of the specific restriction site. This is equivalent to the usual Michaelis-Menten equation except that V_{max} and K_m have both been reduced by a factor $(K_i + lK_m)/K_i$. Hence linear Lineweaver-Burk plots will be obtained but the measured parameters will depend on the values of K_m, K_i and l which in general will vary from DNA to DNA.

E Immobilized enzymes

An enzyme immobilized on a solid support should have advantages over one in free solution:

(1) it can easily be separated from the products and unreacted substrate and hence can be used many times;
(2) its thermal stability should be increased.

It may suffer two disadvantages:

(i) the enzyme's activity may be adversely affected by the chemical reaction involved in coupling it to the inert support;
(ii) the rate of diffusion of substrate into and product out of the active site may be greatly reduced.

Both Bam HI and Eco RI have been coupled to Sepharose 4B by the cyanogen bromide activation method (Lee *et al.*, 1978). The enzymes hydrolyse λ DNA, adenovirus DNA and SV40 DNA to give the same products as does the enzyme in free solution. As expected the immobilized enzymes are considerably more stable. Bam HI can be heated to 65°C without significant loss of activity and the matrix bound enzymes can also be lyophilized and stored without loss of activity.

F Recognition sites

The overwhelming majority of restriction enzymes recognize a particular tetra or hexanucleotide sequence. In a random DNA (which will of course contain 50% A—T base pairs (bp), 50% G—C bp a particular tetranucleotide sequence will occur on average every 256 bp, a hexanucleotide will be expected every 4096 bp. Most DNA is not of course "random" and this leads to some interesting observations. In eukaryotic DNA for example there is a deficiency of the dinucleotide sequence CpG (Morrison *et al.*, 1967). This results in there being only one site for Hpa I (recognition sequence CCGG) in SV40 DNA and no sites at all for Tha I (CGCG) whereas the similar sized prokaryotic DNA from the plasmid pBR322 contains 26 and 23 sites respectively (Table 2).

In a DNA which is 75% A—T, 25% G—C, the recognition site for Sma I (CCCGGG) should occur only once every 262 144 bp. In fact in the DNA from *Euglena gracilis* chloroplast (130 kilobase pairs, kb) there are no Sma I sites at all (Gray and Hallick, 1978).

The significance of all the above is that judicious choice of restriction enzyme will fragment DNA into pieces sufficiently large to

Table 2 Some restriction sites containing the sequence CpG.

Hha I	GCGC
Hpa II	CCGG
Taq I	TCGA
Tha I	CGCG
Ava I	GCCGGC
Hae II	AGCGCT
Hind II	GTCGAC
Sal I	GTCGAC
Sma I	CCCGGG
Xho I	CTCGAG
Xma I	CCCGGG

contain the entire coding sequence for most proteins together with any introns or control sequences. Suitable combinations of restriction enzymes can be used to produce fragments as small as desired.

Many recognition sites contain a two-fold axis of symmetry. For example Eco RI recognizes

5′ . . . . GAATTC . . . . 3′

3′ . . . . CTTAAG . . . . 5′

The symmetrical tetranucleotides (of which there are 16 sequences) and the symmetrical hexanucleotides (of which there are 64 sequences) can be displayed in the form of a chart (Fig. 5).

This chart reveals at a glance that there is a strong tendency for restriction enzyme sites to be GC-rich. The reasons for this are not clear but of course such features influence the expected frequency of sites as discussed above. The possibility (Gierer, 1966) that such symmetrical recognition sequences might allow the formation of short cruciform structures has been much debated (Malcolm, 1977). No conclusive evidence for these structures in such short sequences has been obtained although in longer sequences they seem highly probable (Lilley, 1980; Panayotatos and Wells, 1981).

Another useful feature of the restriction enzymes hydrolysis of DNA is the position of bonds which are attacked. Some enzymes cut the DNA in such a way as to produce termini which are fully base paired (or flush ended). For example Hae III cuts as shown

5′ . . . . G . p . G$^{\downarrow}$p . C . p . C . . . . 3′

3′ . . . . C . p . C . p$_{\uparrow}$ G . p . G . . . . 5′

However a substantial number of enzymes cut to produce termini with a single stranded end (usually 2 or 4 bases). The cleavage by

5′ → 3′	Tetra	A . . . . T
AATT	EcoRI*(4)	
ATAT		
ACGT		
AGCT	AluI (0) OxaI	BbrI ( ) ChuI ( ) Hin91R ( ) Hin173 ( ) HinbIII ( ) HindIII (4) HinFII ( ) HsuI (4)
TATA		
TTAA		
TCGA	TaqI (2)	ClaI (2)
TGCA		AvaIII
CATG		
CTAG		
CCGG	HapII (2) HpaII (2) MnoI (2) MspI (2) SfaGUI ( )	
CGCG	BceR ( ) AccII ( ) FnuDII (0) Hin1056I ThaI (0)	
GATC	DpnI (0) DpnII ( ) FnuAII ( ) FnuCI (4) MboI (4) MosI ( ) Sau 3A (4) FnuEI (4)	BglII (4) XhoII (4)
GTAC	RsaI (0)	RruI ( )
GCGC	CfoI (2) FnuDIII (2) HhaI (2) HinGUI ( ) MnnIV	HaeII (4) HinHI ( ) NgoI ( )
GGCC	BluII ( ) BspRI ( ) BsuRI (0) Bsu1076 Bsu1114 ( ) FnuDI (0) HaeIII (0) HhgI MnnII ( ) NgoII ( ) PalI ( ) SfaI (0)	HaeI (0) GdiI ( ) StuI ( )

Figure 5 Symmetrical restriction enzyme recognition sites. Column 1 gives the central tetranucleotide. Column 2 lists the enzymes recognizing the tetranucleotide. Columns 3–6 list enzymes which cleave hexanucleotides of which

T....A	C....G	G....C
		EcoRI (4) RsrI
		AcyI (2) AosII (2)
	PvuII (0)	HgiAI (4) SacI (4) SstI (4)
		AccI (2) SnaI
		HpaI (0) ChuII () HincII ( ) HindII (0) MnnI
	BluI (4) ScuI ( ) SexI ( ) SgoI ( ) SlaI (4) SluI ( ) SpaI ( ) Xho (4) XpaI (4) Xho (4) XpaI (4) AvaI (4) AvrI ( ) AquI ( )	SalI (4) XamI ( ) AccI (2) ChuII ( ) HincII ( ) HindIII (0) MnnI ( )
		HgiAI (4)
XbaI (4)	AvrII ( )	
	SmaI (0) XmaI (4) AvaI (4) AvrI ( ) AquI	
	SacII (2) SstII ( ) TglI ( )	
AtuCI ( ) BclI (4) CpeI ( ) SstIV	PvuI ( ) RshI (2) XniI (0) XorII (4)	BamFI ( ) BamHI ( ) BamKI ( ) BamNI BstI (4) XhoII (4)
		KpnI (4)
AosI ( ) MstI ( )		AcyI (2) AosII (2) HaeII (4) HinHI NgoI ( )
HaeI (0) BalI (0)	XmaIII (4)	

the first and last nucleotides are given at the top of the column and the central tetranucleotide is that given in Column 1. Based on a suggestion by Dr. L. W. Coggins. The number in parenthesis indicates the length of the projecting single strand region after cleavage.

Table 3 Restriction enzymes which produce single stranded projecting ends.

5′ extensions		
	Apy I	CC↓$^{A}_{T}$GG
	Fnu 4HI	GC↓NGC
	Acc I	GT$^{CT}_{AG}$AC
	Acy I	GPu↓CGPyC
	Asu II	TT↓CGAC
	Cla I	AT↓CGAT
	Hpa II	C↓CGG
	Taq I	T↓CGA
	Ava II	G↓G$^{A}_{T}$CC
	Dde I	C↓TNAG
	Hinf I	G↓ANTC
	Sau 96I	GG↓NCC
	Ava I	C↓PyCGPuG
	Bam HI	G↓GATCC
	Bcl I	T↓GATCA
	Bgl II	A↓GATCT
	Eco RI	G↓AATTC
	Hind III	A↓AGCTT
	Hsu I	A↓AGCTT
	Sal I	G↓TCGAC
	Sau 3A	GATC
	Xba I	T↓CTAGA
	Xho I	C↓TCGAG
	Xho II	Pu↓GATCPy
	Xma I	C↓CCGGG
	Xma III	C↓GGCCG
	Bst EII	G↓GTNACC
	Eco RII	↓CC$^{A}_{T}$GG
3′ extensions		
	Hha I	GCG↓C
	Pvu I	CGAT↓CG
	Sst II	CCGC↓GG
	Bgl I	GCCNNNN↓NGGC
	Hae II	PuGCGC↓Py
	Hgi AI	G$^{A}_{T}$GC$^{A}_{T}$↓C
	Kpn I	GGTA↓CC
	Pst I	CTGCA↓G
	Sph I	GCATG↓C
	Sst I	GAGCT↓C
	Xor II	CGATC↓G

Eco RI produces termini whose structures are

5′ . . . . G   pApApTpTC . . . .

3′ . . . . CpTpTpApAp   G . . . .

A consequence of this is that any Eco RI-produced terminus will be able to base pair with any other Eco RI end. A suitable DNA ligase can then be used to repair the two phosphodiester bonds and two pre-

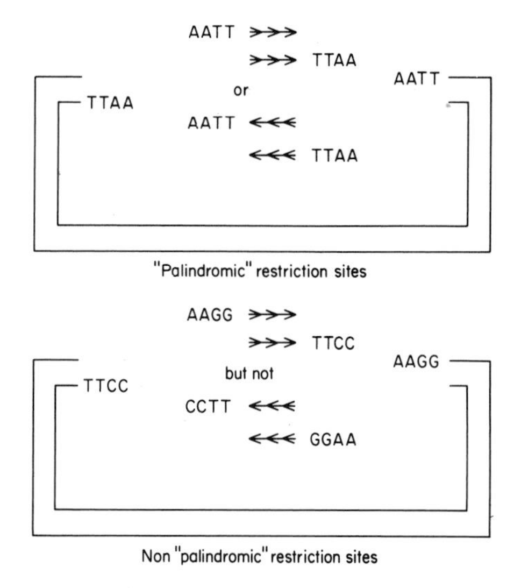

Figure 6 A comparison of religation occurring between symmetrical and non-symmetrical restriction enzyme generated termini.

viously unrelated DNA fragments have now formed a "recombinant" DNA molecule.

Table 3 shows the more widely used enzymes arranged according to the length of the unpaired end and whether it is a 3′ or 5′ extension.

Note that the 3′ end is always the hydroxyl while the 5′ end carries the phosphate. It is worth emphasizing that although the enzymes Bam HI, Bcl I, Bgl II and Xho II recognize different hexanucleotides, they all produce the same (GATC) single strand extension. Similarly, Sal I, Xho I and Ava I all leave TCGA as a projecting 5′ terminus.

From the point of view of constructing recombinant DNA molecules it is in fact a disadvantage that so many restriction enzyme sites are symmetrical because this means that fragments can recombine in either orientation (Fig. 6). Since expression of genes recombined in this way will normally be critically dependent on the orientation, this is an important point.

Table 4 lists the enzymes which will produce fragments where the ends may not all be the same.

This point is perhaps best illustrated by the example of Hga I acting on ϕ X174 DNA. Hga I cuts

$$5' \ldots\ldots \text{GACGC } (\text{N})_5{\downarrow}$$

$$3' \ldots\ldots \text{CTGCG } (\text{N})_5 \ (\text{N})_5{\uparrow}$$

Table 4 Restriction enzymes which leave non-identical ends after cleavage.

Mnl I	CCTC $(N)_5$
Eco PI	AGACC
Atu I	CC↓$^{A}_{T}$GG
Eco RII	↓CC$^{A}_{T}$GG
Dde I	C↓TNAG
Mbo II	GAAGA
Hga I	GACGC
Sfa NI	GATGC
Hinf I	G↓ANTC
Bbv I	GC$^{A}_{T}$GC
Fnu 4HI	GCNGC
Hph I	GGTGA
Ava II	G↓G$^{A}_{T}$CC
Sau 96I	GGNCC
Bst EII	G↓GTNACC
Bgl I	GCC $(N)_4$NGGC

(Brown and Smith, 1977) and there are 4^5 (= 1024) different possible projecting ends. The 14 fragments actually produced by Hga I digestion of ϕX174 are only capable of recombining in the correct order, and in fact ligation of the fragments does produce infective ϕX174 DNA.

G Other nucleic acid substrates

In addition to cleaving double stranded DNA, several restriction enzymes are able to cleave other types of nucleic acids.

Molloy and Symons (1980) have shown that the following enzymes will cleave DNA/RNA hybrids: Eco RI, Hind II, Sal I, Msp I, Hha I, Alu I, Taq I and Hae III. Such cleavage could be of use for providing restriction maps of RNA molecules (since an RNA molecule can be used as a template for DNA synthesis in a reverse transcriptase catalysed reaction).

Several restriction enzymes appear to cut single stranded DNA. Hae III, Hha I and Sfa I hydrolyse single stranded ϕX174 and M13 DNA (Blakesley and Wells, 1975). Hae III will also cut ss fI DNA (Horiuchi and Zinder, 1975) but in both cases hydrolysis is much slower than with a double stranded template. Godson and Roberts (1976) showed that, in addition to the above, Mbo II, Hinf I, Hpa II, Pst I, Blu I and Ava I all hydrolyse single stranded ϕX174 and other phage DNAs.

This ability of Hae III and Hha I to cleave single stranded DNA was used by Seeburg *et al.* (1981) in order to help characterize the complementary DNA for rat growth hormone.

There is however a question as to whether the enzymes are really cleaving single stranded DNA. Blakesley *et al.* (1977) have used ligands such as actinomycin D and netropsin which should only bind to double stranded DNA, to demonstrate that within a molecule which is regarded as single stranded overall, such as ϕX(+) DNA, there are regions where the conformation of the molecule allows double helical regions to form and they suggest that it is these regions which act as substrates for restriction enzyme cleavage.

Finally an intriguing discovery was made recently by Bishop (1979). The plasmid pCM21 contains the sequence

$$5' \ldots . \text{G}\downarrow\text{AATTA} \ldots .$$
$$3' \ldots . \text{C} \quad \text{TTAAT} \ldots .$$

and this is cleaved by Eco RI in one strand only, at the position shown by the arrow. This suggests that at least for Eco RI, cleavage occurs independently in the strands and this is supported by the elegant kinetic studies of Halford *et al.* (1979).

H Specificity

Most restriction enzymes are assayed or used in a buffer close to 100 mM Tris HCl pH 7.5, 10–50 mM NaCl, 5 mM $MgCl_2$, 5 mM 2-mercaptoethanol, 1 mM EDTA. Temperatures of 37° C and incubation times of 1 h to 15 h are commonly used.

There are several points worth making about these conditions.

1 "Star" activity

Several enzymes have been found to have a specificity which depends on the precise assay conditions used. Eco RI cleaves GAATTC at pH 7.3 and 100 mM NaCl in the presence of 5 mM $MgCl_2$ but raising the pH or lowering the NaCl concentration or substituting Mn^{2+} for Mg^{2+} or adding organic solvents tends to reduce the specificity to cleavage at the more common sequence AATT (Polisky *et al.*, 1975; Hsu and Berg, 1978; Mayer, 1978; Woodhead *et al.*, 1981). Other enzymes which have been discovered to have a variable specificity include Dde I (Makula and Meagher, 1980), Bst I (Clarke and Hartley, 1979), Bam HI (George *et al.*, 1980), Bsu I (Heininger *et al.*, 1977), Xba I, Sal I, Hha I, Pst I and Sst I (Malyguine *et al.*, 1980). In few of these cases has it been established what the new sequence specificity is — the evidence is based on an increase in the number of bands seen in a gel after a limit digestion.

2 *Temperature*

Several restriction enzymes have been isolated from thermophilic organisms: Tth III I (Shinomiya and Sato, 1980); Tth III II (Shinomiya *et al.*, 1980); Taq I (Sato *et al.*, 1977); Bst I (Clarke and Hartley, 1979); Tha I (McConnell *et al.*, 1978). Many of these will therefore work quite satisfactorily at temperatures of 60° C or 70° C which has the advantage that most contaminating nucleases will be denatured. Also of course the restriction will be several times faster than at 37° C. The only possible disadvantage (and this may be only theoretical) is that at temperatures as high as 70° C there may be conformational changes in the DNA which will affect the ability of the enzyme to digest it. For example, Eco RI will not digest the octanucleotide TGAATTCA above 17° C — the temperature at which it melts (Greene *et al.*, 1975). Short double stranded sequences are sometimes not cleaved — for example the decanucleotide CCAAGCTTGG which contains the recognition sequences for Hsu I and Hind III (AAGCTT) is not cut by either of them (Scheller *et al.*, 1977; Miller *et al.*, 1980). There are two Mbo I sites close to each other in polyomavirus DNA (positions 1333 and 1338). Either of these can be cut by Mbo I but once one has been cut, the other is extremely resistant to digestion (P. Deininger, personal communication). The assumption in both cases is that the enzyme requires a flanking sequence of some sort perhaps to prevent the DNA "fraying at the ends" and producing single stranded nucleic acid. Such fraying at the ends has been shown to occur by NMR (Patel, 1980).

I Molecular properties

It is perhaps fortunate that a detailed knowledge of the molecular properties of these enzymes is of little importance to their use in genetic engineering since very little is known. Several seem to be dimers of two identical subunits each of molecular weight around 30 000 daltons (Modrich, 1979). This is hardly surprising of view of the symmetry properties of many of the recognition sites.

The primary sequence of Eco RI has been determined by sequencing the DNA of the gene coding for it (Young *et al.*, 1981). Preliminary X-ray data have also been published for Eco RI (Young *et al.*, 1981). Restriction enzymes almost certainly bind non-specifically to DNA as well as to their recognition sequence (Woodhead and Malcolm, 1980a) and this will usually result in inhibition of activity (Hinsch *et al.*, 1980; Langowski *et al.*, 1980) as has been previously discussed.

Again, not surprisingly, the positively charged side chains of lysine and arginine seem to be important for the hydrolytic activity and/or

DNA binding (Lee and Chirikjian, 1979; Woodhead and Malcolm, 1980a) of at least two enzymes — Bgl I and Eco RI. The activity of Eco RI depends on a uniquely reactive carboxyl side chain (Woodhead and Malcolm, 1980b) and this is interesting in view of the strong interaction between guanine and carboxyl groups (Lancelot and Hélène, 1977).

The effect of alteration of the bases in the recognition site for Eco RI has also been studied. The sequence IAATTC with inosine replacing guanine is cleaved as rapidly as is GAATTC (Modrich and Rubin, 1977). GAAUUC is also cleaved but hydroxymethylation of either U or C leads to a substantial reduction in hydrolysis (Berkner and Folk, 1977).

The fact that the rates at different sites for an enzyme vary so much suggests that sequences around the recognition site are also involved in interactions with the enzyme. This may be exploited to produce inhibition at selected sites by the use of sequence specific ligands (such as actinomycin D which binds preferentially to . . . . GpC . . . . or netropsin which prefers . . . . ApApA . . . .) (Braga *et al.*, 1975; Nosikov *et al.*, 1976; Fania and Fanning, 1976; Cohen *et al.*, 1980). This trick has been exploited by Nosikov *et al.* (1978) to obtain a restriction map of bacteriophage λ DNA with the enzyme Hpa I. A recent paper has shown how rates at individual sites may be measured and this has allowed such observations to be quantitated (Malcolm and Moffatt, 1981).

III Analytical applications

The applications of restriction enzymes to genetic engineering may be crudely divided into two sections: the analytical and the synthetic.

A Counting gene copies

Restriction enzymes may be used to count the number of copies of a gene sequence (or of closely related, cross hybridizing, ones). For example it is known that there is no site for either of the restriction enzymes Taq I or Hind III within the coding sequence for human leukocyte interferon. Restriction of total DNA by either of these enzymes followed by electrophoresis, transfer to cellulose nitrate and hybridization to mRNA shows the presence of eight bands suggesting that there is at least this number of genes (Goeddel *et al.*, 1981). A similar approach shows that there is only one copy of the gene for the β subunit of human chorionic gonadotropin (Fiddes and Goodman, 1979).

B Restriction mapping

Restriction enzymes have been very widely used to map small genomes such as plasmids, bacteriophages and viruses, to compare different strains of these and to produce short fragments for sequencing.

Mapping is fairly simple. The DNA of interest is digested with an enzyme, the fragments electrophoresed through agarose or polyacrylamide and their molecular weights determined. It is possible to align these fragments in their correct order by electron microscopy (Keegstra *et al.*, 1977; Miwa *et al.*, 1979) but it is more usual to use analysis of partial digests, digestion by mixtures of enzymes or by end labelling the DNA prior to further digestion.

The ends of a DNA restriction fragment may be labelled in a variety of ways of which the two most widely used are:

1 *"Klenow" labelling*

The large fragment produced by limited proteolysis of DNA polymerase retains the $5' \rightarrow 3'$ synthetic activity (Setlow *et al.*, 1972) and can be used to fill in the projecting 5′ single stranded terminus produced by many enzymes (Table 3)(Sanger *et al.*, 1977):

$$\begin{array}{lcl} & [\alpha^{32}P]\ \text{dATP} & \\ & [\alpha^{32}P]\ \text{dTTP} & \\ 5' \ldots\ldots \text{G} & & 5' \ldots\ldots \text{G}^{32}\text{pA}^{32}\text{pA}^{32}\text{pT}^{32}\text{pT} \\ & \longrightarrow & \\ 3' \ldots\ldots \text{CpTpTpApAp} & & 3' \ldots\ldots \text{C pT pT pA pA} \\ & \text{"Klenow"} & \\ & \text{enzyme} & \end{array}$$

2 *Kinase labelling*

The phosphoryl group at the 5′ terminus of a DNA strand may be removed by treatment with a phosphatase enzyme. Treatment with $[\gamma^{32}P]$ ATP in the presence of polynucleotide kinase will result in the restoration of the 5′ phosphoryl group but now it will be radioactive (Chaconas and Van De Sande, 1980).

An example will illustrate this method. The plasmid pBR322 is cut twice by Hinc II to give fragments of 3255 bp and 1106 bp. The ends produced may be labelled either by kinase or by "Klenow" enzyme. Subsequent cleavage by Eco RI (which has only a single

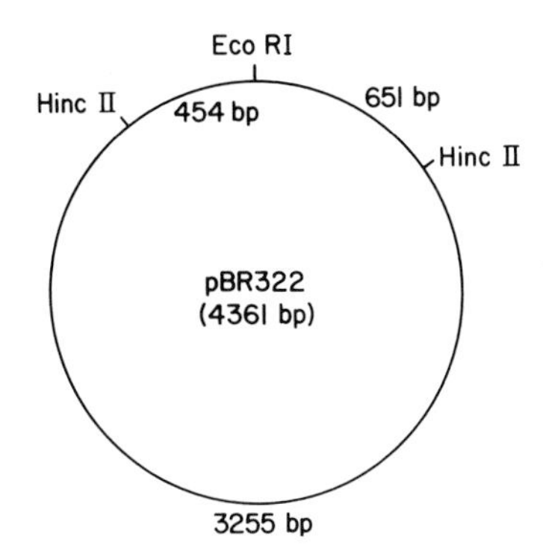

Figure 7 Mapping the Eco RI and Hinc II restriction sites in pBR322.

site in pBR322) results in the appearance of fragments of 3255 bp, 651 bp and 454 bp and this allows the positions of the two Hinc II sites to be oriented with respect to the Eco RI site as shown in Fig. 7.

There are two elegant extensions of methods which take advantage of the non-identity of the ends produced by some enzymes.

Buluwela *et al.* (1981) have used the enzyme Dde I which hydrolyses the sequence

$$\begin{array}{l} 5' \ldots . \mathrm{C}^{\downarrow}\,\mathrm{TNA}\;\;\mathrm{G} \\ 3' \ldots . \mathrm{G}\;\;\mathrm{AMT}_{\uparrow}\mathrm{C}. \end{array}$$

Four different ends are possible depending on the nature of N and this means that careful selection of [α^{32}P] NTPs for the Klenow labelling reaction can produce ten different possible ways in which a fragment could be labelled. Analysis of the actual patterns enables the fragments to be aligned along the genome more readily.

A variation on this has been developed by Baralle *et al.* (1980) using Hinf I which cleaves at

$$\begin{array}{l} \mathrm{G}^{\downarrow}\,\mathrm{ANT}\;\;\mathrm{C} \\ \mathrm{C}\;\;\mathrm{TMA}_{\uparrow}\mathrm{G} \end{array}$$

These authors have studied the region around the human globin genes. A 2.0 kb Bam HI/Eco RI fragment was treated with Hinf I and the projecting ends "filled in" using the "Klenow" fragment and [α^{32}P] dATP as the only radioactive nucleotide. These DNA fragments were then denatured and hybridized to M13 ϵ 3.7 + DNA. (This is a single stranded bacteriophage DNA molecule containing an inserted 3 700 bp of the ϵ-globin gene region (Baralle *et al.*, 1980). This hybrid molecule was then again filled in with the "Klenow" fragment. Redigestion with Hinf I results in the radioactive label appearing in the fragment adjacent to that in which it originated

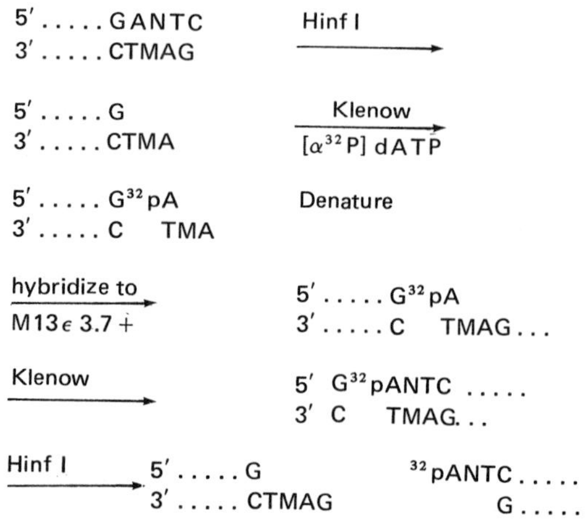

```
5'.....GANTC          Hinf I
3'.....CTMAG          ------->

5'.....G                Klenow
3'.....CTMA           ------->
                      [α32P] dATP

5'.....G32pA          Denature
3'.....C   TMA

hybridize to          5'.....G32pA
------->              3'.....C   TMAG...
M13ε 3.7 +

Klenow                5' G32pANTC .....
------->              3' C   TMAG...

Hinf I         5'.....G              32pANTC.....
------->       3'.....CTMAG                G.....
```

Figure 8 The use of Hinf I for ordering restriction fragments.

(Fig. 8). This enables the restriction fragments to be aligned along the genome.

C Sequencing

After the restriction fragments have been positioned it is now a routine matter to sequence any one of interest. Standard techniques exist for eluting the desired fragment from a gel (Smith, 1980), for end labelling the fragment (see Chs 6–10, Grossman and Moldave, 1980), and for sequencing (see Chs 56–65, Grossman and Moldave, 1980). Normally a length of around 100 bases will be sequenced on a single gel. This implies that the fragment in question will be the product of a double or even triple digest, since cleavage with a single enzyme will give fragments much bigger than this.

D Screening recombinants

If it is wished to demonstrate the presence of a particular inserted sequence in a recombinant molecule, restriction enzyme analysis is ideal. For example the plasmid pBR322 contains a single Eco RI site. Insertion of a β-globin cDNA sequence at the Pst I site would result in the appearance of a second Eco RI site, since the globin DNA also possesses one site (spanning the positions corresponding to amino acid residues 121/122).

A particularly elegant application of this technique has been used by Fiddes and Goodman (1980) in their search for a cloned cDNA corresponding to the β subunit of human chorionic gonadotropin (β HCG). The restriction enzymes Asu I and Sau 96I both recognize the sequence GGNCC. Now all codons beginning GG specify glycine, while those starting CC code for proline and these are the *only* codons for these two amino acids. Thus wherever the dipeptide Gly—Pro occurs in a protein, the DNA is bound to contain a restriction site for Asu I and Sau 96I. A search of the known amino acid sequence for β HCG showed that Gly—Pro occurred at positions 102/103 and at 136/137 and therefore the cDNA must contain at least two Sau 96I sites 102 bp apart. As it happens residues 104—106 (Lys Asp His) and 122—124 (Lys Ala Pro) could be encoded by AAGGACCNN and by AAGGCCCNN respectively providing that the appropriate codons are used. Two of the recombinants proved to have Sau 96I fragments 54 and 39 bp long suggesting that these recombinants must contain the β HCG coding sequence and that the codon utilization is indeed that suggested above.

As with several other nucleases, restriction enzymes may be used to study the binding of other proteins to DNA. Thus one of the four Hpa I sites in SV40 DNA becomes resistant to digestion in the presence of *E. coli* RNA polymerase (Allet *et al.*, 1974).

E DNA methylation

It has been known for many years that a small proportion (2—10%) of DNA is methylated (on the *N* position of adenine or the 5 position of cytosine). In higher animals it is only the cytosine residues which are methylated. It is probably fair to say that at the time of writing, the exact significance of this is not yet clear (Burdon and Adams, 1980) but there is circumstantial evidence linking it to differentiation and control of gene expression. Restriction enzymes are of course ideal tools with which to study methylation because it is methylation of its own DNA by the original host which is thought to confer protection against restriction. Thus Hha I cleaves unmethylated GCGC sequences and Hpa II cleaves unmethylated CCGG but neither will cleave if the cytosines have been methylated. The enzyme Msp I is both interesting and useful since it cleaves CCGG whether or not the cytosines are methylated (Waalwijk and Flavell, 1978), and thus a comparison between the digestion patterns of Hpa II and Msp I enables the number and location of methylated cytosines to be determined. Mammalian sperm DNA, for example, has been shown to be very highly methylated (Waalwijk and Flavell, 1978).

It does however appear as though methylation reduces the rate of digestion by Msp I (Sneider, 1980).

The methylation of adenines may be studied in a similar way by using the trio Sau 3A (cleaves both GATC and G ^{Me}ATC), Dpn I (cleaves only G ^{Me}ATC) and Mbo I (cleaves only GATC).

F Evolution

The earliest molecular analyses of evolution were of course based on the amino acid sequences of proteins and from these some conclusions could be made about the likely base changes involved. Because of the quantities of material required such studies referred to species rather than to individuals within species. Also of course it was not possible in this way to study the DNA which did not express itself in protein.

Analysis of restriction enzyme sites allows both these disadvantages to be overcome. Jeffreys (1979) has studied the occurrence of eight different restriction sites in the $\gamma\delta\beta$-globin gene locus in 60 unrelated humans. Three different variant restriction sites were detected — one Pst I and two Hind III sites were involved — and none occurred within a section of DNA coding for protein. On the basis of a statistical analysis of the data, Jeffreys concludes that on average at least 1 in 100 bp along the human genome will vary polymorphically in man.

G Genetic defects

Several genetic diseases are known to be caused by simple mutations or deletions in the DNA of a structural gene. For example a change from CTC to CAC at the sixth codon in β-globin changes the amino acid from glutamate to valine giving rise to sickle cell disease in homozygotes. Another example is where a change in the codon at position 17 of the β-globin gene from CTT to CTA changes the amino acid specified from lysine to a stop signal causing premature termination of translation giving rise to β thalassaemia.

Clearly if such a mutation either removes or produces a restriction enzyme site then this can be used as a marker for the presence or absence of the defect. Such information can of course be used to design an antenatal diagnosis for the defect (Marx, 1978). Thus the sickle cell mutation causes the loss of a Dde I (CTNAG) site and the HbO Arab mutation a loss of the only Eco RI (GAATTC) site in the β chain of globin.

As it happens it is not even necessary that the genetic defect itself should involve a change of restriction site. A restriction enzyme site polymorphism sufficiently closely linked to the defect is quite sufficient. For example in people with HbA (the normal adult haemoglobin) the gene coding for the β-globin polypeptide chain is located within a Hpa I fragment 7.6 kb or 7.0 kb in length (Kan and Dozy, 1978a). However in 87% of persons with HbS (sickle cell haemoglobin), this fragment is 13.0 kb long as a result of a second mutation which leads to the loss of one Hpa I recognition site. This polymorphism is so closely linked to the actual genetic defect that the chance of recombination occurring between these two sites is small. This loss of a Hpa I site may therefore be regarded as diagnostic for the disease and has been used in antenatal detection (Kan and Dozy, 1978b).

For more information on this topic the reader is referred to the excellent review in Volume 1 of this series (Little, 1981).

IV Synthetic applications

A Modifying ends

As already mentioned any end of a DNA fragment which has been generated by, say, Bam HI will be complementary to any other end generated by the same enzyme. It will therefore be easy to make a recombinant molecule between two fragments each generated by the same enzyme. In general however it will be necessary to be able to unite molecules generated by different restriction enzymes, or even, in the case of cDNA, by no restriction enzyme.

1 Removal of phosphate

The phosphoryl moiety on the 5′ hydroxyl of any piece of DNA can be removed using alkaline phosphatase. Treatment of the potential vector in this way has one great advantage — it prevents the original DNA from recombining with itself to reform circular DNA. This helps to reduce the background of antibiotic resistant plasmids which do not contain an inserted sequence.

2 Addition of phosphate

A phosphoryl group can be added to the 5′ hydroxyl group by the use of T4 polynucleotide kinase. The extent of labelling can readily be followed if [$\gamma^{32}P$] ATP is used as the phosphoryl donor.

3 Removal of projecting ends

If the two restriction ends to be joined are not in fact compatible then one approach is to remove the projecting ends and produce a blunt end (i.e. fully base paired) fragment. This may be done by using a single strand specific nuclease such as that from *Aspergillus oryzae*, known as Sl (Shishido and Ando, 1972).

4 *Filling in projecting ends*

As already mentioned (see Section III.A.1), mild proteolysis of DNA polymerase I results in its cleavage into two fragments, the larger of which retains the original $5' \rightarrow 3'$ polymerizing activity but lacks the $5' \rightarrow 3'$ exonuclease activity (Klenow *et al.*, 1971; Setlow *et al.*, 1972). This fragment (known as the "Klenow" enzyme or A fragment) can therefore be used to fill in a projecting $5'$ terminus produced by restriction cleavage (Fig. 8).

As can be seen from Table 3 the majority of projecting ends are of this type. There is no corresponding method for filling in projecting $3'$ termini.

5 *Projecting ends from blunt ends*

Blunt ends can be converted into single strand ends in a variety of ways.

The exonuclease from bacteriophage λ (Little, 1967) will hydrolyse one strand of double stranded DNA starting at the $5'$ end. Exonuclease III from *E. coli* (Richardson *et al.*, 1964) will perform the same function but starts from the $3'$ end. In practice neither of these is widely used since the resulting single stranded end is not necessarily complementary to any other.

6 *Terminal transferase*

A much more widely used approach involves terminal deoxynucleotidyl transferase which catalyses the addition of nucleotides to the $3'$ end of a polynucleotide:

$$d(pY)_m + n dXTP \rightarrow d(pY)_m\ (pX)_m + n PP_i$$

This reaction can therefore be used to add, say, ten CMP residues to one of the DNA molecules to be joined and ten GMP residues to the other, thus producing complementary ends with the capability of forming a large number of hydrogen bonds. The average length of the "tails" can readily be monitored by following the uptake of [^{32}P] when [α^{32}P] NTPs are used as substrates. It should be noted that when purine residues are being added, the best divalent metal ion to use as co-factor is Mg^{2+}, whereas with pyrimidines Co^{2+} is the ion of choice (Kato *et al.*, 1967).

This enzyme can also be used to introduce [^{32}P] at the $3'$ end of a DNA strand (cf. Section IV.A.2). Terminal transferase will catalyse the addition of ribonucleotides as well as deoxyribonucleotides. Subsequent treatment with alkali will break the phosphate ester bonds just formed except for the one attached to the $3'$ hydroxyl at the end of the DNA chain, since this residue lacks a 2′ hydroxyl (a prerequisite for alkali lability). If the original nucleoside triphosphate was α labelled then this ^{32}P will be left behind on the DNA (Chang *et al.*, 1977).

B Synthetic linkers

Advances in nucleotide chemistry during the last five years (Khorana, 1979) and in particular the development of the triester method of polynucleotide synthesis (Itakura *et al.*, 1975), has meant that oligonucleotides carrying particular restriction sites are readily available. A wide variety is available commercially. For example CCGAATTCGG/GGCTTAAGCC contains the Eco RI recognition site, CCGGATCCGG/GGCCTAGGCC includes recognition sequences for Bam HI and Hpa II, CCAAGCTTGG/GGTTCGAACC contains both Hind III and Alu I sites (Scheller *et al.*, 1977). Such "linkers", being self complementary, will spontaneously form base paired double helical structures and can readily be ligated to an existing "flush" end, and then be converted into the cohesive end corresponding to the artificially added restriction enzyme.

Table 5 shows the range of linkers currently available commercially together with the enzyme sites they contain.

Longer ones are available and the list is continually being extended. All the various molecules produced as a result of the manipulations detailed in Sections IV.A and B can be joined by the use of phage T4 DNA ligase (the gene for which has recently been cloned).

C Site directed mutagenesis

Restriction enzymes can be used to introduce single strand "nicks" at specific points in a DNA sequence (see Section II.G). If the nicked DNA is now treated with DNA polymerase I in the presence of deoxynucleotide triphosphates then the DNA will undergo "nick translation" (Kelly *et al.*, 1970). If this is carried out in the presence of a nucleotide analogue then this may be incorporated instead of the correct base. If the nick translation is carried out in the absence of one or more of the nucleoside triphosphates then the extent of the incorporation of the analogue will be limited to one or two positions near the site of the original restriction enzyme attack. An example will illustrate the approach.

Weissmann *et al.*, (1979) took the plasmid pβG which contains cDNA corresponding to most of β-globin.

This insert contains an Eco RI site at positions corresponding to amino acids 121/122. The introduction of a single strand nick at this point followed by nick translation using dATP and hydroxy dCTP

Table 5 Some commercially available restriction enzyme "linkers".

AATT	Eco RI*
CCGG	Hpa II, Msp I
GGCC	Hae III
AGCT	Alu I
GCGC	Hha I
CGCG	Tha I
ATGCAT	Ava III
TCTAGA	Xba I
GTTAAC	Hpa I
GGAATTCC	Eco RI
CAAGCTTG	Hind III
CGGATCCG	Bam HI
GCTCGAC	Taq I, Ava I, Xho I
CTGATCAG	Bcl I
CAGATCTG	Bgl II
CATCGATG	Cla I, Taq I
CGGTACCG	Kpn I
GCTGCAGC	Pst I
CGAGCTCG	Sst I, Sac I
GGTCGACC	Sal I
CTCTAGAG	Xba I
CCTCGAGG	Xho I
CCGAATTCGG	Eco RI
CCAAGCTTGG	Hind III
CCGGATCCGG	Bam HI

results in the incorporation of the analogue close to the Eco RI site. The important point however is the HOdC residues can base pair with A and in the absence of dTTP will be incorporated in its place. However when the substituted plasmid is replicated *in vivo* in the presence of sufficient concentrations of all four deoxynucleoside triphosphates then dG will be incorporated opposite the HOdC and thus the overall result will be the substitution of a G—C base pair in place of an A—T base pair.

In spite of its elegance, this trick has not been greatly exploited, probably because of the development of a method which can be used even where there is no convenient restriction site. In this a short oligonucleotide (> eight residues) complementary to the region of interest except for a single base change is synthesized. This is then hybridized to the DNA sequence and used as a primer for DNA polymerase I. This results in a complete complementary sequence being synthesized except for the single base change at the position of choice.

For example the correct sequence of ϕX174 DNA from 582—593 reads TTTGTAGGATAC and the synthetic sequence GTATCCCACAAA will hybridize to this and give rise to a change from A → G at position 587 after replication (Hutchinson *et al.*, 1978). Such an approach can also be used to correct mutations (Razin *et al.*, 1978) and to produce mutations in the TATA box region which is thought to be important for transcription by eukaryotic RNA polymerase II (Wasylyk *et al.*, 1980).

D Examples

A few examples may help to illustrate the use of restriction enzymes in the synthetic aspects of genetic engineering.

1 Somatostatin

This is a polypeptide hormone whose action is antagonistic to that of growth hormone. It is only 14 amino acids long and it was possible to predict a DNA sequence which could code for this (owing to the redundancy of the genetic code, there is of course no guarantee that it is the same as the actual sequence used *in vivo*). This sequence was synthesized by the triester method and a suitable vector designed to maximize the chance of expression was prepared (Itakura *et al.*, 1977).

The plasmid pBR322 (see Section V.A for its restriction map) was cleaved at its single Eco RI site. The projecting ends were filled in and ligated to a 203 bp Hae III fragment containing the lac promoter, the CAP binding site, the operator, ribosome binding site and the first seven codons for β-galactosidase. An important feature of this step is that EcoRI sites are regenerated at each end of this insert:

```
—G                          AATTC—        (Eco RI-cleaved pBR322)
—CTTAA                          G—
                ↓     Fill in
—GAATT                     AATTC—
—CTTAA                     TTAAG—
                ↓     Add      CC—lac—GG
                               GG—lac—CC
                            + ligase
—GAATTCC —    lac    — GGAATTC—
—CTTAAGG — system — CCTTAAG —
```

This plasmid is called pBH10 and is screened by virtue of its resistance both to tetracycline and to ampicillin and its possession of a pair of Eco RI sites 200 bp apart (Section III.B).

One of these sites is sufficiently close to the promoter for the tetracycline resistance gene that *E. coli* RNA polymerase binding will protect it from Eco RI attack. Digestion of the other Eco RI site, followed by removal of the single strand tail by S1 nuclease (Section IV.A.3) and religation produces plasmid pBH20 which differs from pBH10 only in the loss of this Eco RI site.

pBH20 was then treated with Eco RI and Bam HI which removed about 300 bp. Removal of the terminal phosphates (Section IV.A.1) then prevented recircularization, and the linear molecule was ligated to the synthetic somatostatin coding sequence. This plasmid, pSomI, could be screened by its ampicillin resistance, tetracycline sensitivity and its possession of an Eco RI/Bam HI fragment of the correct size.

Unfortunately although this plasmid now contained the somatostatin coding sequence preceded by the lac control region, no hormone could be detected in these cells by standard radioimmune assay.

In order to increase the stability of the polypeptide a similar series of manipulations was undertaken with the end result that the β galactosidase structural gene was inserted between the lac control region and the somatostatin gene. The protein product (which could be induced by isopropyl thiogalactoside) consists of a hybrid between β-galactosidase and somatostatin. Since the latter contains no internal methionine residues, treatment of this molecule with cyanogen bromide releases immunoreactive somatostatin.

2 *Insulin*

This polypeptide hormone, synthesized in the pancreas, controls the uptake of glucose into cells. Its two polypeptide chains (A and B) are in fact derived from a single precursor (proinsulin) by the proteolytic removal of a short connecting sequence (the C peptide).

The gene for rat proinsulin I (the rat has two genes for this hormone) was cloned by isolating mRNA, producing a cDNA copy and removing the single stranded regions with Sl nuclease (Section IV.A.3). The synthetic decanucleotide $\genfrac{}{}{0pt}{}{\mathrm{CCAAGCTTGG}}{\mathrm{GGTTCGAACC}}$ which contains a recognition site for HindIII was then ligated to each end of the cDNA and treated with HindIII to produce cohesive projecting ends. The plasmid pMB9 has a single HindIII site very close to the promoter for the tetracycline resistance gene and this plasmid was treated with HindIII and then with alkaline phosphatase (Section IV.A.1) to prevent the parent pMB9 from recyclizing without the insert. The cDNA was then ligated to the pMB9 (Ullrich *et al.*, 1977).

A similar approach has been used to close the human insulin gene (Bell *et al.*, 1979) except that this time the cDNA was inserted into the PstI site of pBR322 (Section V.B) using homopolymer tailing

(Section IV.A.5). The various colonies obtained in this experiment were screened by hybridization to the rat cDNA clone described above, since the sequences are sufficiently similar that cross hybridization occurs.

The human insulin cDNA clone was then used as a probe to screen a human genomic DNA library for the actual human insulin gene (Bell *et al.*, 1980). A comparison of the restriction enzyme maps of the genomic clone with the cDNA clone immediately reveals the presence of an intervening sequence (intron) since the total extent of the coding region is considerably greater in the genomic clone than in the cDNA clone. Also the former is cleaved by Sma I to give two fragments each of which hybridizes to a cDNA probe, although the cDNA itself does not contain a site for Sma I.

Coding sequences for the individual A and B chains have also been synthesized by the triester method. These have been inserted together with the lac operon between the Eco RI and Bam HI sites of pBR322 to produce ampicillin resistant, tetracycline sensitive colonies. The proteins produced consists of hybrids between β-galactosidase and one or other of the insulin chains (Riggs *et al.*, 1979). Insulin synthesized from isolated chains prepared in this way appears to be clinically useful.

E Maximizing expression of a cloned gene

The greatest uses of cloned sequences will come if in addition to being replicated in bacteria they can also be expressed in bacteria. T.M. Roberts *et al.* (1979) have reported a method in which the DNA sequence being studied is inserted into a plasmid with its 5′ end about 100 base pairs from a unique restriction endonuclease site. The plasmid is then opened and the distance between this opening and the end of the gene is gradually reduced by a combination of exonuclease III and SI treatment (Sections IV.A.3 and 5). The lac promoter fragment previously mentioned (Section IV.D.1) is then added to produce a collection of clones where the lac promoter is positioned at varying distances from the 5′ end of the gene to be expressed. Provided that screening for the desired product is possible then a suitable clone should be readily detected. The authors used the cro protein of bacteriophage λ as their example but there seems no reason why eukaryotic cDNA sequences should not be expressed in exactly the same way.

V Restriction maps

The bacteriophage λ, the plasmid pBR322 and the animal virus SV40 are probably the most used vectors and for convenience this section describes the restriction maps of these three DNAs. It should be noted that various (defective) strains of the former are widely used, and that the complete DNA sequences of pBR322 (Sutcliffe, 1979) and SV40 (Fiers *et al.*, 1978) have been published.

A Bacteriophage λ

This is a linear molecule of total length 49 kb. The positions of the restriction sites are given in map units where the total length of the

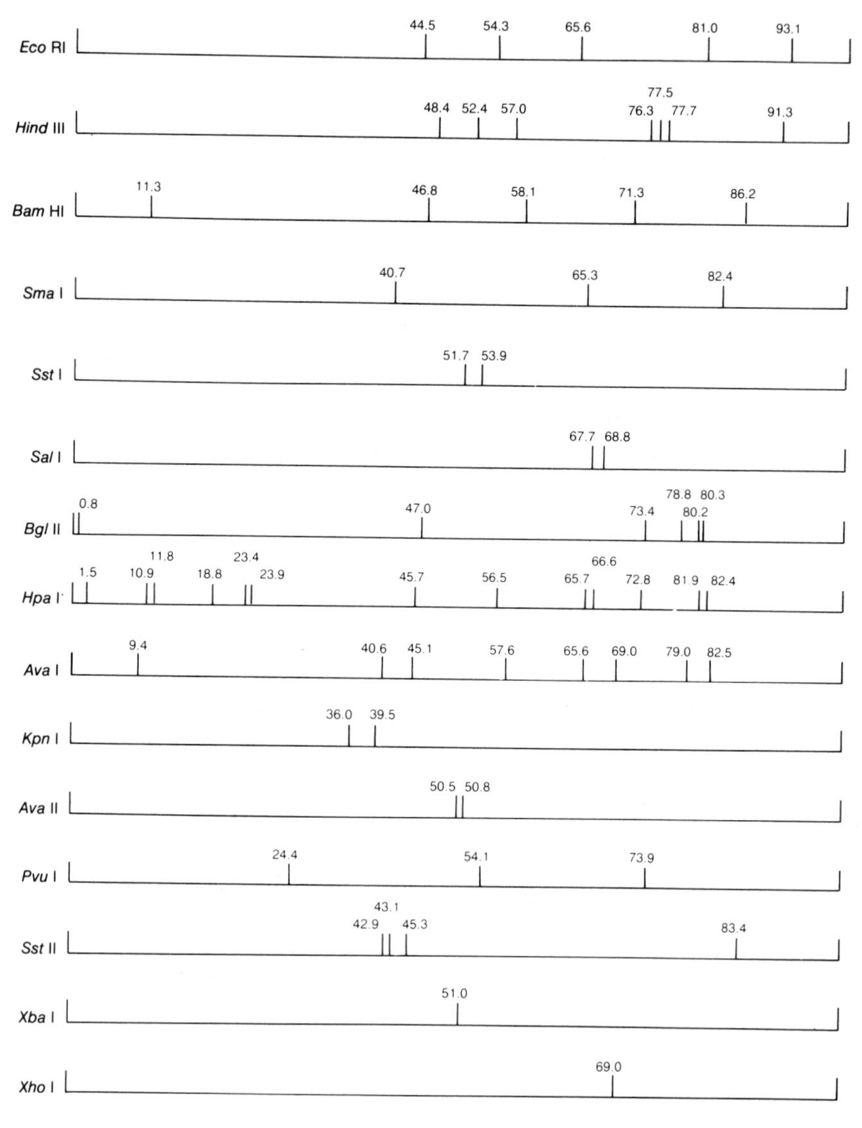

Figure 9 Restriction map of bacteriophage λ.

molecule is equal to 100 (Szybalski and Szybalski, 1979). The fragment sizes (which are often used as molecular weight markers) are:

Eco RI	21.8 kb	7.55 kb	5.9 kb	5.5 kb	4.8 kb	3.4 kb		
Hind III	23.7 kb	9.5 kb	6.7 kb	4.3 kb	2.25 kb	1.95 kb	0.6 kb	0.1 kb
Bam HI	17.4 kb	7.3 kb	6.8 kb	6.5 kb	5.5 kb	5.5 kb		
Sma I	19.9 kb	12.1 kb	8.62 kb	8.4 kb				
Sst I	25.3 kb	22.6 kb	1.1 kb					
Sal I	33.2 kb	15.3 kb	0.5 kb					

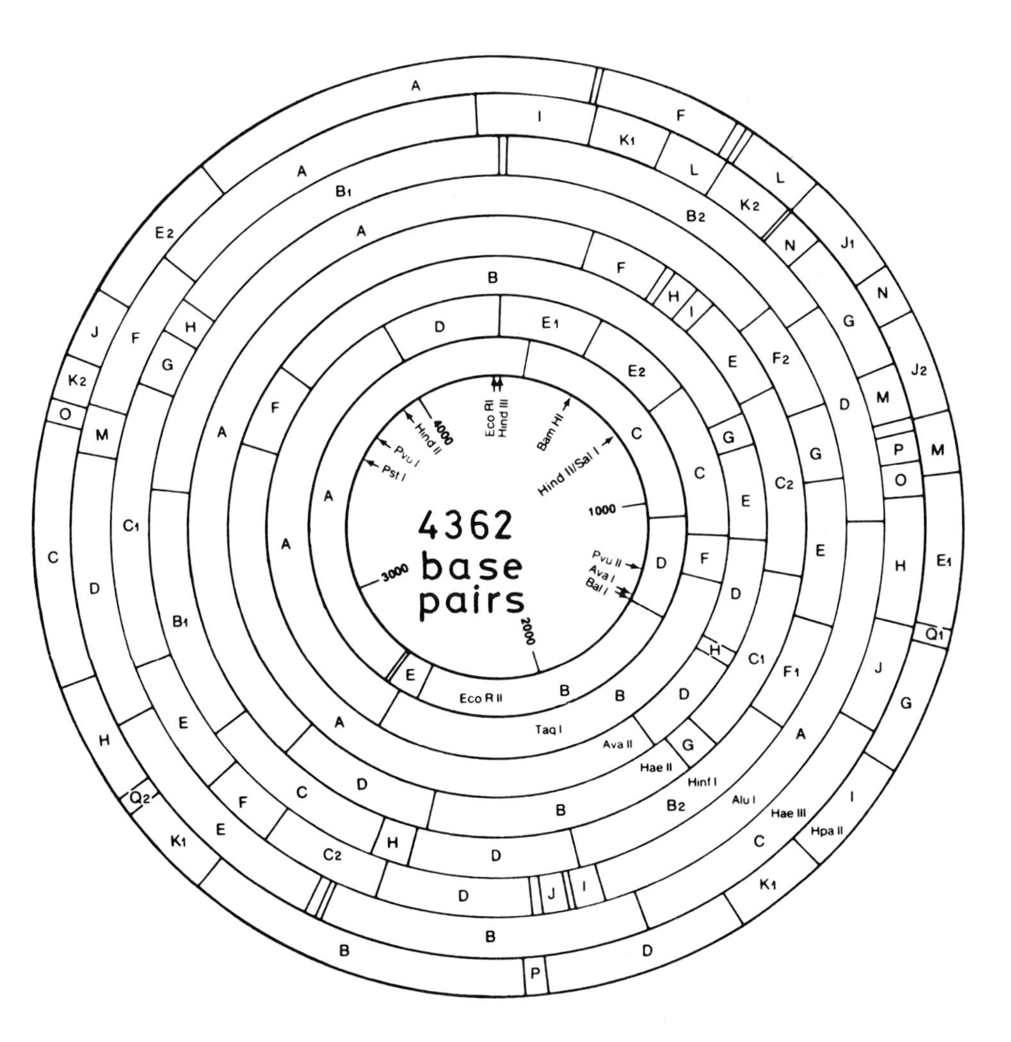

Figure 10 Restriction map of pBR322.

B pBR322

There are no sites for Bcl I, Bgl II, Hpa I, Kpn I, Sma I, Sst I, Xba I or Xho I in pBR322. (See opposite.)

C SV40

There are no sites in SV40 for Ava I, Bal I, Bgl II, Bln I, Hga I, Sal I, Sma I and Xba I.

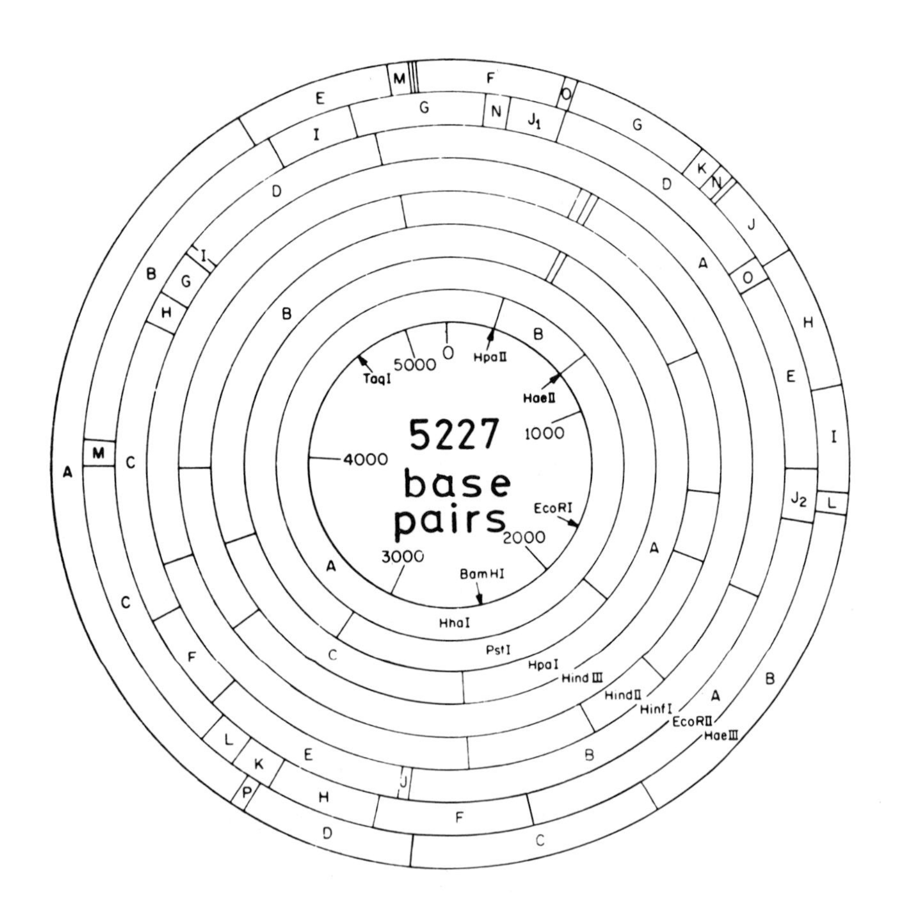

Figure 11 Restriction map of SV40 DNA.

VI Commercial suppliers

This list is intended to be neither exhaustive nor a recommendation. It is simply the names and addresses of companies with whom the author has dealt:

Bethesda Research Laboratories
411 North Stonestreet Avenue
PO Box 6010, Rockville
Maryland 20850, USA

Boehringer Mannheim GmbH Biochemica
PO Box 31 01 20
6800 Mannheim 31
West Germany

Collaborative Research Inc.
1365 Main Street, Waltham
Massachusetts 02154, USA

Miles Research Products
PO Box 37, Stoke Poges
Slough, Berkshire SL2 4LY
UK

New England Biolabs
283 Cabot Street, Beverly
Massachusetts 01915, USA

P-L. Biochemicals Inc.
1037 West McKinley Avenue
Milwaukee, Wisconsin 53205, USA

VII Acknowledgements

I am grateful to many colleagues for help and advice and in particular to J.R. Moffatt and J.L. Woodhead without whose help my interest in restriction endonucleases would have remained theoretical. Work on restriction enzymes in my laboratory has been supported by the Cancer Research Campaign, the Science Research Council and the Wellcome Trust.

VIII References

Allet, B., Roberts, R. J., Gesteland, R. F. and Solem, R. (1974). *Nature, Lond.* **249**, 217–220.

Arber, W. (1971). In "The Bacteriophage λ" (A. D. Hershey, ed.), 83–96. Cold Spring Harbor Laboratory, New York.

Arber, W. (1974). *Prog. Nucl. Acids Res. Mol. Biol.* **14**, 1–37.

Arrand, J. R., Myers, P. A. and Roberts, R. J. (1978). *J. Mol. Biol.* **118**, 127–135.

Bahl, C. P., Wu, R., Brousseau, R., Sood, A. K., Hsiung, H. M. and Narang, S. A. (1978). *Biochem. Biophys. Res. Commun.* **81**, 695–703.

Baksi, K. and Rushizky, G. W. (1979) *Analyt. Biochem.* **99**, 207–212.
Baksi, K., Rogerson, D. L. and Rushizky, G. W. (1978). *Biochemistry* **17**, 4136–4139.
Baralle, F. E., Shoulders, C. C., Goodbourne, S., Jeffreys, A. and Proudfoot, N. J. (1980a) *Nucl. Acids Res.* **8**, 4393–4404.
Baralle, F. E., Shoulders, C. C. and Proudfoot, N. J. (1980b). *Cell* **21**, 621–626.
Bell, G. I., Swain, W. F., Pictet, R., Cordell, B., Goodman, H. M. and Rutter, W. J. (1979). *Nature, Lond.* **282**, 525–527.
Bell, G. I., Pictet, R. L., Rutter, W. J., Cordell, B., Tischer, E. and Goodman, H. M. (1980). *Nature, Lond.* **284**, 26–32.
Berkner, K. L. and Folk, W. R. (1977). *J. Biol. Chem.* **252**, 3185–3193.
Bertani, G. and Weigle, J. J. (1953). *J. Bacteriol.* **65**, 113.
Bickle, T. A., Pirrotta, V. and Imber, R. (1977). *Nucl. Acids Res.* **4**, 2561–2572.
Bickle, T. A., Pirrotta, V. and Imber, R. (1980). In "Methods in Enzymology" (L. Grossman and K. Moldave, eds) Vol. 65, 132–138. Academic Press, New York.
Bingham, A. H. A., Atkinson, T., Sciaky, D.and Roberts, R. J. (1978). *Nucl. Acids Res.* **5**, 3457–3467.
Bishop, J. O. (1979). *J. Mol. Biol.* **128**, 545–559.
Bron, S. and Hörz, W. (1980). In "Methods in Enzymology", (L Grossman and K. Moldave, eds) Vol. 65, 112–132. Academic Press, New York.
Blakesley, R. W. and Wells, R. D. (1975). *Nature, Lond.* **257**, 421–422.
Blakesley, R. W., Dodgson, J. B., Nes, I. F. and Wells, R. D. (1977). *J. Biol. Chem.* **252**, 7300–7306.
Braga, E. A., Nosikov, V. V., Tangashin, V. I., Zhuze, A. L. and Polyanovskij, O. L. (1975). *Dokl. Akad. Nauk SSSR* **225**, 707–710.
Brown, N. L. and Smith, M. (1977). Proc. Natn. Acad. Sci. U.S.A. **74**, 3213–3216.
Brown, N. L., McClelland, M. and Whitehead, P. R. (1980). *Gene* **9**, 49–68.
Buluwela, L., Malcolm, A. D. B., Coleman, D. V. and Gardner, S. D. (1981). *Bioscience Rep.* **1**, 223–228.
Burdon, R. H. and Adams R. L. P. (1980). *Trends in Biochem Sci.* **5**, 294–297.
Burgess, R. R. and Jendrisak, J. J. (1975). *Biochemistry* **14**, 4634–4638.
Catterall, J. F. and Welker, N. E. (1980). In "Methods in Enzymology" (L. Grossman and K. Moldave, eds) Vol. 65, 167–170. Academic Press, New York.
Chaconas, G. and Van De Sande, J. H. (1980). In "Methods in Enzymology" (L. Grossman and K. Moldave, eds) Vol. 65, 75–85. Academic Press, New York.
Chang, J. C., Temple, G. F., Poon, R., Neumann, K. H. and Kan, Y. W. (1977). *Proc. Natn. Acad. Sci. U.S.A.* **74**, 5145–5149.
Chater, K. (1977). *Nucl. Acids Res.* **4**, 1989–1998.
Clanton, D. J., Woodward, J. M. and Miller, R. V. (1978). *J. Bacteriol.* **135**, 270–273.
Clarke, C. M. and Hartley, B. S. (1979) *Biochem. J.* **177**, 49–62.
Cohen, G. L., Ledner, J. A., Bauer, W. R., Ushay, H. M., Caravana, C. and Lippard, S. J. (1980). *J. Am. Chem. Soc.* **102**, 2487–2488.
DeWaard, A., Korsuize, J., Van Beveren, C. P. and Maat, J. (1978). *FEBS Letters* **96**, 106–110.
DeWaard, A., Van Beveren, C. P., Duyvesteyn, M. and Van Ormondt, H. (1979). *FEBS Letters* **101**, 71–76.
Duncan, C. H., Wilson, G. A. and Young, F. E. (1978). *J. Bacteriol.* **134**, 338–344.

Duyvesteyn, M. and DeWaard, A. (1980). *FEBS Letters* **111**, 423–426.
Endow, S. A. and Roberts, R. J. (1977). *J. Mol. Biol.* **112**, 521–529.
Fania, J. and Fanning, T. G. (1976). **67**, 367–371.
Fiddes, J. C. and Goodman, H. M. (1979). *Nature, Lond.* **281**, 351–356.
Fiddes, J. C. and Goodman, H. M. (1980). *Nature, Lond.* **286**, 684–687.
Fiers, W. *et al.* (1978). *Nature, Lond.* **273**, 113–120.
Fuchs, L. Y., Covarubbias, L., Escalante, L., Sanchez, S. and Bolivar, F. (1980). *Gene* **10**, 39–46.
Gelinas, R. E. (1977). *J. Mol. Biol.* **114**, 433–440.
Gelinas, R. E., Myers, P. A. and Roberts, R. J. (1977). *J. Mol. Biol.* **114**, 169–180.
George, J., Blakesley, R. W. and Chirikjian, J. G. (1980). *J. Biol. Chem.* **255**, 6521–6524.
Gierer, A. (1966). *Nature, Lond.* **212**, 1480–1481.
Gingeras, T. R., Myers, P. A., Olson, J. A., Hanberg, F. A. and Roberts, R. J. (1978). *J. Mol. Biol.* **118**, 113–122.
Godson, G. N. and Roberts, R. J. (1976). *Virology* **73**, 561–567.
Goeddel, D. V., Leung, D. W., Dull, T. J. and Gross, M. (1981). *Nature, Lond.* **290**, 20–26.
Goff, S. and Rambach, A. (1978). *Gene* **3**, 347–352.
Graham, A., Greene, J., Lowe, P. A. and Malcolm, A. D. B. (1976). *Biochem. Soc. Trans.* **4**, 633–634.
Gray, P. W. and Hallick, R. B. (1978). *Biochemistry* **17**, 284–289.
Greenaway, P. J. (1980). *Biochem. Biophys. Res. Commun.* **95**, 1282–1287.
Greene, P. J., Betlach, M. C., Goodman, H. M. and Boyer, H. W. (1974). *Meth. Mol. Biol.* **7**, 87–111.
Greene, P. J., Poonian, M. S., Nussbaum, A. L., Tobias, L., Garfin, D. E. and Boyer, H. W. (1975). *J. Mol. Biol.* **99**, 237–261.
Grossman, L. and Moldave, K. (Eds) (1980). "Methods in Enzymology", Vol. 65. Academic Press, New York.
Halford, S. E., Johnson, N. P., and Grinsted, J. (1979). *Biochem, J.* **179**, 353–365.
Hartmann, H. and Goebel, W. (1977). *Febs Letters* **80**, 285–287.
Heininger, K., Hort, W. and Zachau, H. G. (1977). *Gene* **1**. 291–303.
Hines, J. L., Chauncey, T. R. and Agarwal, K. L. (1980). In "Methods in Enzymology" (L. Grossman and K. Moldave, eds). Vol. 65, 152–163. Academic Press, New York.
Hinkle, N. F. and Milter, R. V. (1979). *Plasmid* **2**, 387–393.
Hinsch, B., Mayer, H. and Kula, M-R. (1980). *Nucl. Acids Res.* **8**, 2547–2559.
Hobom, G., Schwarz, E., Melzer, M. and Mayer, H. (1981). *Nucl. Acids Res.* **9**, i–xxx.
Horiuchi, K. and Zinder, N. D. (1975). *Proc. Natn. Acad. Sci. U.S.A.* **72**, 2555–2558.
Hsu, M. and Berg, P. (1978). *Biochemistry* **17**, 131–138.
Hu, A. W., Kuebbing, D. and Blakesley, R. (1978). *Fedn Proc.* **37**, 1415.
Hughes, S. G., Bruce, T. and Murray, K. (1980). *Biochem. J.* **185**, 59–63.
Hutchinson III, C. A., Phillips, S., Edgell, M. H., Gillam, S., Jahnke, P., and Smith, M. (1978). *J. Biol. Chem.* **253**, 6551–6560.
Ikawa, S., Shibata, T. and Ando, T. (1976). *J. Biochem.* **80**, 1457–1460.
Itakura, K., Katagiri, N., Bahl, C. P., Wightman, R. and Narang, S. A. (1975). *J. Am. Chem. Soc.* **97**, 7327–7331.
Itakura, K., Hirose, T., Crea, R., Riggs, A. D., Heyneker, H. L., Bolivar, F. and Boyer, H. W. (1977). *Science* **198**, 1056–1063.

Jeffreys, A. J. (1979). *Cell* **18**, 1–10.
Jeppesen, P. G. N. (1980). In "Methods in Enzymology" (L. Grossmann and K. Moldave, eds) Vol. 65, 305–319. Academic Press, New York.
Kan, Y. W. and Dozy, A. M. (1978a). *Proc. Natn. Acad. Sci. U.S.A.* **75**, 5631–5635.
Kan, Y. W. and Dozy, A. M. (1978b). *Lancet* ii, 910–911.
Kato, K-I., Gonsalves, T. M., Houts, G. E. and Bollum, F. J. (1967). *J. Biol. Chem.* **242**, 2780–2789.
Kauc, L. and Piekarowicz, A. (1978). *Eur. J. Biochem.* **92**, 417–426.
Keegstra, W., Vereijken, J. M. and Jansz, H. S. (1977). *Biochim. Biophys. Acta* **475**, 176–183.
Kelly, R. B., Cozzarelli, N. R., Deutscher, M. P., Lehman, I. R. and Kornberg, A. (1970). *J. Biol. Chem.* **245**, 39–45.
Kelly, T. J. and Smith, H. O. (1970). *J. Mol. Biol.* **51**, 393–409.
Khorana, H. G. (1979). *Science* **203**, 614–625.
Kiss, A., Sain, B., Csordas-Toth, E. and Venetianer, P. (1977). *Gene* **1**, 323–329.
Kleid, D. G. (1980). In "Methods in Enzymology" (L. Grossmann and K. Moldave, eds) Vol. 65, 163–166. Academic Press, New York.
Klenow, H., Overgaard-Hansen, K. and Patkar, S. A. (1971). *Eur. J. Biochem.* **22**, 371–381.
Kunkel, L. M., Silberklang, M. and McCarthy, B. J. (1979). *J. Mol. Biol.* **132**, 133–139.
Lacks, S. A. (1980). In "Methods in Enzymology" (L. Grossmann and K. Moldave, eds) Vol. 65, 138–146. Acadamic Press, New York.
Lancelot, G. and Hélène, C. (1977). *Proc. Natn. Acad. Sci. U.S.A.* **74**, 4872–4875.
Landy, A., Ruedisueli, E., Robinson, L. Foeller, C. and Ross, W. (1974). *Biochemistry* **13**, 2134–2142.
Langowski, J., Pingoud, A. Goppelt, M. and Maass, G., (1980). *Nucl. Acids Res.* **8**, 4727–4736.
Lebon, J. M., Kado, C., Rosenthal, L. J. and Chirikjian, J. G. (1978). *Proc. Natn. Acad. Sci. U.S.A.* **75**, 4097–4101.
Lee, Y-H. and Chirikjian, J. G. (1979). *J. Biol. Chem.* **254**, 6838–6841.
Lee, Y-H., Blakesley, R. W., Smith, L. A. and Chirikjian, J. G. (1978). *Nucl. Acids Res.* **5**, 679–689.
Leung, D. W., Lui, A. C. P., Meritees, H., McBride, B. C. and Smith, M. (1979). *Nucl. Acids Res.* **6**, 17–25.
Lilley, D. M. J. (1980). *Proc. Natn. Acad. Sci. U.S.A.* **77**, 6468–6472.
Little, J. W. (1967). *J. Biol. Chem.* **242**, 679–686.
Little, P. F. R. (1981). "Genetic Engineering 1 (R. Williamson, ed.). Academic Press, London and New York. 61–102.
Lui, A. C. P., McBride, B. C., Vovis, G. F. and Smith, M. (1979). *Nucl. Acids Res.* **6**, 1–15.
Lynn, S. P., Cohen, L. K., Gardner, J. F. and Kaplan, S. (1979). *J. Bacteriol.* **138**, 505–509.
Lynn, S. P., Cohen, L. K., Kaplan, S. and Gardner, J. F. (1980). *J. Bacteriol.* **142**, 380–383.
McConnell, D., Searcy, D. and Sutcliffe, G. (1978). *Nucl. Acids Res.* **5**, 1729–1739.
Makula, R. A. and Meagher, R. B. (1980). *Nucl. Acids Res.* **8**, 3125, 3131.
Malcolm, A. D. B. (1977). *Nature, Lond.* **268**, 196–197.
Malcolm, A. D. B. and Moffatt, J. R. (1981). *Biochim. Biophys. Acta*, in press.

Malyguine, E., Vannier, P. and Yot, P. (1980). *Gene* **8**, 163–177.
Mann, M. B., Rao, R. N. and Smith, H. O. (1978). *Gene* **3**, 97–112.
Marx, J. L. (1978). *Science* **202**, 1068–1069.
Mayer, H. (1978). *FEBS Letters* **90**, 341–344.
Meselson, M. and Yuan, R. (1968). *Nature, Lond.* **217**, 110–1114.
Messelson, M., Yuan, R. and Heywood, J. (1972). *A. Rev. Biochem.* **41**, 447–460.
Middleton, J. H., Edgell, M. H. and Hutchinson, C. A. (1972). *J. Virol.* **10**, 42–50.
Miller, P. S., Cheung, D. M., Dreon, N. Jayamaran, K., Kan, L-S. Leutzinger, E. E. and Pulford, S. M. (1980). *Biochemistry*, **19**, 4688–4698.
Miwa, T., Takanami, M. and Yamagishi, H. (1979). *Gene* **6**, 319–330.
Modrich, P. (1979). *Q. Rev. Biophys.* **12**, 315–369.
Modrich, P. and Rubin, R. A. (1977). *J. Biol. Chem.* **252**, 7273–7278.
Molloy, P. L. and Symons, R. H. (1980). *Nucl. Acids Res.* **8**, 2939–2946.
Murray, K., Hughes, S. G., Brown, J. S. and Bruce, S. (1976). *Biochem. J.* **159**, 317–322.
Nathans, D. and Smith, H. O. (1975). *A. Rev. Biochem.* **44**, 273–293.
Nosikov, V. V., Braga, E. A., Karishev, A. V. and Zhuze, A. L. (1976). *Nucl. Acids Res.* **3**, 2293–2301.
Nosikov, V. V., Braga, E. A. and Sain, B. (1978). *Gene* **4**, 69–84.
Old, R., Murray, K. and Roizes, G. (1975). *J. Mol. Biol.* **92**, 331–339.
Panayotatos, N. and Wells, R. D. (1981). *Nature, Lond.* **289**, 466–470.
Patal, D. J. (1980). Proc. Br. Biophys. Soc. Abstract.
Pirrotta, V. and Bickle, T. A. (1980). In "Methods in Enzymology" (L. Grossman and K. Moldave, eds) Vol. 65, 89–95. Academic Press, New York.
Polisky, B., Greene, P. Garfin, D. E., McCarthy, B. J., Goodman, H. M. and Boyer, H. W. (1975). *Proc. Natn. Acad. Sci. U.S.A.* **72**, 3310–3314.
Pugatsch, T. and Weber, H. (1979). *Nucl. Acids Res.* **7**, 1429–1444.
Rambach, A. (1980). In "Methods in Enzymology" (L. Grossman and K. Moldave, eds) Vol. 65, 170–173. Academic Press, New York.
Razin, A., Hirose, T., Itakura, K. and Riggs, A. D. (1978). *Proc. Natn. Acad. Sci. U.S.A.* **75**, 4268–4270.;
Richardson, C. C., Lehman, I. R. and Kornberg, A. (1964). *J. Biol. Chem.* **239**, 251–258.
Riggs, A. D., Itakura, K., Crea, R., Hirose, T., Kraszewski, A., Bolivar, F. Heyneker, H., Kleid, D. and Goeddel, D. (1979). XIth International Congress of Biochemistry (Toronto). Abstracts, p. 76.
Roberts, R. J. (1976). *Crit. Rev. Biochem.* **4**, 123–164.
Roberts, R. J. (1981). *Nucl. Acids Res.* **9**, 75–96.
Roberts, R. J., Breitmeyer, J. B., Tabachnik, N. F. and Myers, P. A. (1975). *J. Mol. Biol.* **91**, 121–123.
Roberts, R. J., Myers, P. A., Morrison, A. and Murray, K. (1976). *J. Mol. Biol.* **102**, 157–165.
Roberts, T. M., Kacich, R. and Ptashne, M. (1979). *Proc. Natn. Acad. Sci. U.S.A.* **76**, 760–764.
Roizes, G., Patillon, M. and Kovoor, A. (1977). *FEBS Letters* **82**, 69–70.
Roizes, G., Pages, M., Lecou, C., Patillon, M. and Kovoor, A. (1979). *Gene* **6**, 43–50.
Rubin, R. A. and Modrich, P. (1980). In "Methods in Enzymology", (L. Grossman and K. Moldave, eds) Vol. 65, 96–104. Academic Press, New York.

Sanger, F., Nicklen, S. and Coulson, A. R. (1977). *Proc. Natn. Acad. Sci. U.S.A.* **74**, 5463.
Sato, S., Hutchinson, C. A. and Harris, J. J. (1977). *Proc. Natn. Acad. Sci. U.S.A.* **77**, 542–546.
Sato, S., Nakazawa, K. and Shinomiya, T. (1980). *J. Biochim.* **88**, 737–747.
Scheller, R. H., Dickerson, R. E., Boyer, H. W. and Riggs, A. D. (1977). *Science* **196**, 177–180.
Seeburg, P. H., Shine, J., Martial, J. A., Baxter, J. D. and Goodman, H. M. (1981). *Nature, Lond.* **270**, 486–494.
Setlow, P., Brutlag, D. and Kornberg, A. (1972). *J. Biol. Chem.* **247**, 224–231.
Sharp, P. A., Sugden, B. and Sambrook, J. (1973). *Biochemistry* **12**, 3055–3063.
Shibata, T. and Ando, T. (1975). *Mol. Gen. Genet.* **138**, 269–280.
Shibata, T., Ikawa, S., Kim, C. and Ando, T. (1976). *J. Bacteriol.* **128**, 473–476.
Shimotsu, H., Takahashi, H. and Saito, H. (1980). *Gene* **11**, 219–226.
Shinomiya, T. and Sato, S. (1980). *Nucl. Acids Res.* **8**, 43–56.
Shinomiya, T., Kobayashi, M. and Sato, S. (1980). *Nucl. Acids Res.* **8**, 3275–3285.
Shishido, K. and Ando, T. (1972). *Biochim. Biophys. Acta* **287**, 477–484.
Smith, D. L., Blattner, F. R. and Davies, J. (1976). *Nucl. Acids Res.* **3**, 343–353.
Smith, H. O. (1980). In "Methods in Enzymology" (L. Grossman and K. Moldave) eds) Vol. 65, 371–380. Academic Press, New York.
Smith, H. O. and Nathan S. D. (1973). *J.Mol. Biol.* **81**, 419–423.
Smith, H. O. and Marley, G. M. (1980). In "Methods in Enzymology", (L. Grossman and K. Moldave, eds) Vol. 65, 104–108. Academic Press, New York.
Smith, H. O. and Wilcox, K. W. (1970). *J. Mol. Biol.* **51**, 379–391.
Sneider, T. W. (1980). *Nucl. Acids Res.* **8**, 3829–3840.
Southern, E. (1979). In "Methods in Enzymology" (L. Grossman and K. Moldave) Vol. 68, 152–176.
Sussenbach, J. S., Monfoort, C. H., Schiphof, R. and Stobberingh, E. E. (1976). *Nucl. Acids Res.* **3**, 3193–3202.
Sussenbach, J. S., Steenberg, P. H., Rost, J. A., Van Leeuwen, W. J. and Van Embden, J. D. A. (1978). *Nucl. Acids Res.* **5**, 1153–1163.
Sutcliffe, J. G. (1979). *Cold Spring Harb. Symp. Quant. Biol.* **43**, 77–90.
Szybalski, E. H. and Szybalski, W. (1979). *Gene* **7**, 217–270.
Takahashi, H., Shimizu, M., Saito, H., Ikeda, Y. and Sugisaki, H. (1979) *Gene* **5**, 9–18.
Takanami, M. (1974) *Meth. Mol. Biol.* **7**, 113–133.
Thomas, M. and Davis R. W. (1975). *J. Mol. Biol.* **91**, 315–328.
Thompson, S. T., Cass, K. H. and Stellwagen, E. (1975). *Proc. Natn. Acad. Sci. U.S.A.* **72**, 669–672.
Timko, J., Horwitz, A. H., Zelinka, J. and Wilcox, G. (1981). *J. Bacteriol.* **145**, 873–879.
Ullrich, A., Shihe, R., Chirgwin, J., Pictet, R., Tischer, E., Rutter, W. J. and Goodman, H. M. (1977) *Science* **196**, 1313–1319.
Vanyusin, B. F. and Dobritza, A. P. (1975). *Biochim. Biophys. Acta* **407**, 61–72.
Venetianer, P. (1980) In "Methods in Enzymology" (L. Grossman and K. Moldave, eds) Vol. 65, 109–112. Academic Press, New York.
Waalwijk, C. and Flavell, R. A. (1978). *Nucl. Acids Res.* **5**, 3231–3236.
Walter, F., Hartmann, M. and Roth, M. (1978). Abstract of XIIth FEBS Meeting. Abstract 0648.
Wang, R. Y. H., Shedlarski, J. G., Farber, M. B. and Kuebbing, D. (1980). *Biochim. Biophys. Acta* **606**, 371–385.

Wasylyk, B. et al. (1980). *Proc. Natn. Acad. Sci. U.S.A.* **77**, 7024–7028
Watson, R., Zuker, M., Martin, S. M. and Visentin, L. P. (1980). *FEBS Letters* **118**, 47–50.
Weissmann, C., Weber, H., Taniguchi, T., Müller, W. and Meyer, F. (1979). *Biochem. Soc. Symp.* **44**, 43–55.
Wilson, G. A. and Young, F. E. (1980). In 'Methods in Enzymology" (L. Grossman and K. Moldave, eds) Vol. 65, 147–153. Academic Press, New York.
Woodhead, J. L. and Malcolm, A. D. B. (1980a). *Nucl. Acids Res.* **8**, 389–402.
Woodhead, J. L. Bhave, N. and Malcolm, A.D.B. (1981). *Eur. J. Biochim.* **115**, 293–296.
Wu, R., King, C. and Jay, E. (1978). *Gene* **4**, 329–336.
Young, T-S., Kim., S-H., Modrich, P., Beth, A. and Jay, A. (1981). *J. Mol. Biol.* **145**, 607–610,
Zain, B. S. and Roberts, R. J. (1977). *J. Mol. Biol.* **115**, 249–255.

Gene cloning in yeast

J. D. BEGGS

Cancer Research Campaign, Eukaryotic Molecular Genetics Research Group, Department of Biochemistry, Imperial College, London, UK

I Introduction

Yeasts are unicellular eukaryotic organisms, encompassing a large number of genera and species (Rose and Harrison, 1969). They have been popular organisms for genetic studies because many yeasts can be propagated stably in either the haploid or the diploid state on simple well-defined media, and can be sporulated. The product of meiotic division, or spore tetrad, can be readily micromanipulated using a light microscope, so that each spore may be germinated individually. *Saccharomyces* species (budding yeasts) in particular have been extensively studied genetically, and well characterized mutations are available in most biochemical pathways (Hawthorne and Mortimer, 1978). *Saccharomyces cerevisiae* (Bakers' yeast) is the species most commonly studied in research laboratories and is used in bread making. This chapter describes the use of *Saccharomyces* species (and to a lesser extent the fission yeast *Schizosaccharomyces pombe*) in gene cloning studies.

Methods for gene cloning have been extended to the use of yeast as a host organism mainly with two aims. Firstly, the ability to introduce individual genes into yeast cells, to apply selection by complementation to identify a clone carrying a specific gene and subsequently to isolate the cloned gene, permits the isolation of virtually any yeast gene. The main limitation of the selection is the availability of a suitable yeast mutant strain as recipient. Secondly, yeast cells, being eukaryotic, should be more suitable than bacteria as host cells in which to clone genes from other eukaryotic organisms where functional gene expression is required, and are more easily propagated than higher eukaryotic tissue culture cells. Relatively few eukaryotic genes are expressed in *E. coli* because of the differences between prokaryotic and eukaryotic mechanisms of transcription and translation, and as a result of different types of post-transcriptional and post-translational modifications, especially the splicing of transcripts to remove intervening sequences present in many eukaryotic genes.

This review outlines the procedure of yeast transformation and describes the properties of different types of cloning vectors currently available, giving some examples of their applications. It attempts to indicate the value of these techniques and the extent of the impact which they have had upon the molecular biology of yeast.

II Procedure for yeast transformation

Yeast cells are surrounded by a thick wall which is resistant to relatively severe physical and chemical treatments and inhibits the uptake of many small molecules. To promote DNA uptake yeast cells must be converted to sphaeroplasts by removal of some of the cell wall. The wall is composed mainly of the polysaccharides chitin, mannans and glucans, with protein, lipid and phosphate components (Phaff, 1971). The polysaccharide in the wall can be digested by treatment with mixtures of β-glucanases such as those found in the gut extract of the snail *Helix pomatia* and in certain bacteria. Some commercial preparations of glucanases which are used are: helicase (Industrie Biologique Française, France), β-glucuronidase (Sigma, UK), glusulase (Endo Laboratory, USA), zymolyase (Kirin Brewery Co., Tokyo, Japan).

The procedure for the preparation of yeast sphaeroplasts (Fig. 1)

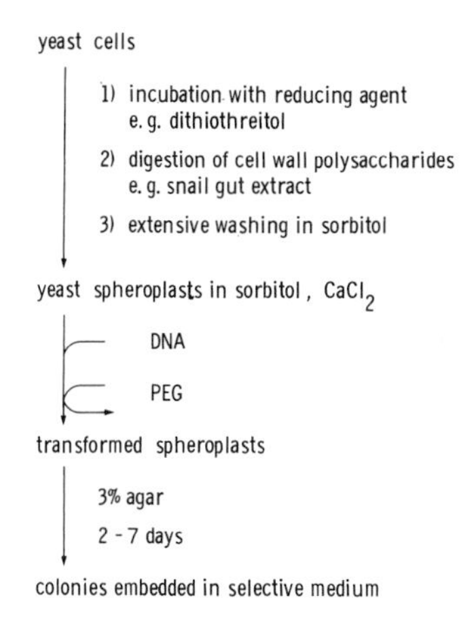

Figure 1 Procedure for transformation of *Saccharomyces cerevisiae.*

is based on that of Van Solingen and Van Der Plaat (1977) developed for sphaeroplast fusion studies. Treatment of the yeast cells with a reducing agent, e.g. β-mercaptoethanol or dithiothreitol, reduces disulphide bridges in the protein components of the wall and sensitizes the wall to the subsequent treatment to hydrolyse the polysaccharides. This prior treatment with a sulphydryl reagent is not always essential and its requirement probably depends on the yeast strain used and the physiological state of the cells to be treated. For example, the walls of yeast cells which have entered the stationary phase of growth are more resistant to digestion than those of cells in mid-log phase. The progress of the sphaeroplasting treatment is usually monitored by testing the sensitivity of the sphaeroplasts to lysis upon dilution in water as opposed to a hypertonic medium (e.g. 1 M sorbitol). The degree of cell lysis can be estimated by measuring the optical density of the sphaeroplast suspensions or by viewing samples with a light microscope. The sphaeroplasts are washed and resuspended in hypertonic medium with calcium chloride and DNA. DNA uptake is promoted by the addition of polyethylene glycol (4000 or 6000 mol. wt; the optimum concentration differs for different sizes and brands of PEG) which also causes the fusion of some of the sphaeroplasts (Van Solingen and Van Der Plaat, 1977). It is not known whether the transient fusion of sphaeroplasts is a necessary intermediate step in DNA uptake; however, some polyploid cells are found among the transformants (Hicks *et al.*, 1979a). The sphaeroplasts are embedded in a solid matrix, usually 3% agar, to facilitate the regeneration of the cell wall (*Schizosaccharomyces pombe* sphaeroplasts will regenerate on the surface of agar; Beach and Nurse, 1981). The sphaeroplasts in molten agar are plated on selective medium which permits only the growth of transformed clones, e.g. a *leu* 2⁻ recipient will only grow on leucine deficient medium if transformed with a *LEU* 2 vector. Colonies develop in the agar in 2—7 days depending upon the particular vector and recipient strain used. A few variations have been developed on this basic method (Hinnen *et al.*, 1978; Beggs, 1978; Struhl *et al.*, 1979) and some vector—recipient strain combinations give better yields of transformants with one or other of the variations (Beggs and Hinnen, unpublished results). It has also been reported (Barney *et al.*, 1980a) that mild treatment of yeast sphaeroplasts with sodium lauryl sarcosinate permits DNA uptake.

III Genetic markers for yeast transformation

Hinnen *et al.* (1978) first demonstrated that yeast could be reproducibly transformed with DNA, that is that yeast cells could take up and maintain exogenously supplied DNA and express genes to alter the phenotype of the recipient cells. They used as donor DNA a recombinant bacterial plasmid (pYe *leu* 2 (10); Ratzkin and Carbon, 1977) carrying the *LEU* 2 gene from *S. cerevisiae.* The *LEU* 2 gene was originally isolated (Ratzkin and Carbon, 1977) in *E. coli* by complementation of a *leu* B6 mutation. The genes *leu* B of *E. coli* and *LEU* 2 of *S. cerevisiae* both code for β-isopropylmalate dehydrogenase, an enzyme required in the biosynthesis of leucine. Several other yeast genes have been cloned by complementation of *E. coli* mutants (Table 1). This is therefore a very simple and convenient method for isolating a few yeast genes whose products, when expressed in *E. coli*, are likely to function to complement the deficiency in the corresponding metabolic pathway of *E coli.* These isolated yeast genes are potential genetic markers for yeast transformation, that is when present in the exogenous DNA their uptake and expression in yeast cells permits the selection of transformed as opposed to untransformed clones.

Some antibiotic resistance determinants of bacterial origin have been shown to be expressed in yeast (Hollenberg, 1979; Cohen *et al.*, 1980). However, since yeast cells are relatively resistant to most of these antibiotics the resistance determinants are not useful as selective markers in transformation. One exception is G418, an aminoglycoside antibiotic which inhibits the growth of a wide range of prokaryotic and eukaryotic organisms, including many yeasts (Jimenez and Davies, 1980). Resistance to G418 is determined by aminoglycoside modifying enzymes one of which is encoded by a bacterial transposable element. Jimenez and Davies (1980)

Table 1 Some of the yeast genes which have been isolated by complementation of mutations in the corresponding *E. coli* genes.

Yeast gene	*E. coli* gene	Gene product	References
HIS 3	*his* B	imidazole glycerol phosphate dehydratase	Struhl *et al.* (1976)
LEU 2	*leu* B	β-isopropylmalate dehydrogenase	Ratzkin and Carbon (1977)
URA 3	*pyr* F	orotidine-5′-phosphate decarboxylase	Bach *et al.* (1979)
TRP 1	*trp* C	*N*-(5′-phosphoribosyl) anthranilate isomerase	Struhl *et al.* (1979)
TRP 5	*trp* A, B	tryptophan synthetase	Walz *et al.* (1978)
ARG 4	*arg* H	argininosuccinate lyase	Clarke and Carbon (1978)

transformed yeast with the transposable element and showed that yeast transformants were resistant to high levels of the antibiotic. This antibiotic resistance determinant should be a useful selectable marker for yeast transformation, especially as the selection should work with any sensitive yeast strain as recipient without the necessity of constructing an appropriate genotype.

In the experiments of Hinnen *et al.* (1978) the plasmid pYe *leu* 2 (10) did not appear to replicate autonomously in the recipient yeast cells. Stable transformants were obtained as a result of recombination between the yeast DNA sequence on the plasmid and the homologous DNA sequence in the nuclear DNA of the recipient cell, such that the pYe *leu* 2 (10) sequence became integrated with chromosome III at the *LEU* 2 locus (or occasionally with another chromosome as a result of a multiple-copy sequence occurring adjacent to *LEU* 2 and also present on other chromosomes).

IV Types of yeast cloning vectors

A Integrating vectors

This type of integrating plasmid can be used as a vector DNA for cloning other DNA sequences in yeast cells. A yeast integrating vector (or YIp) (Struhl *et al.*, 1979) is composed of a bacterial cloning vector (a bacterial plasmid, bacteriophage lambda or cosmid vector) carrying a suitable yeast gene which can be used in the selection of yeast transformants and which provides homology to promote integration of the plasmid DNA into the nuclear DNA of the recipient yeast cell. Transformation efficiencies of 1–10 yeast transformants per microgram of DNA (equivalent to approximately one transformant per 10^6 or 10^7 viable, regenerated yeast cells) are usually obtained. It has been reported (Hicks *et al.*, 1979a, Davis, 1981) that linearization of hybrid plasmids by cleavage within the yeast portion but not within the bacterial portion results in increased transformation frequencies, suggesting that efficient integration of the DNA requires free DNA ends in the region of sequence homology. The transformed phenotype is relatively stable, although after 20 generations of growth approximately 1% of cells will be *leu*$^-$ (see below).

Normally the properties of a DNA cloning vector include:
(1) a good selective marker to identify and purify transformed clones;
(2) unique recognition sites for several type II DNA restriction endonucleases;
(3) the ability to replicate autonomously in the host cell;

(4) easy purification of large amounts of the vector DNA from the host cells;
(5) high efficiency of transformation (many transformants per microgram of vector DNA).

This type of integration vector is therefore unusual in that there will normally be only a single copy of the vector DNA sequence in each stably transformed yeast cell and this copy will only replicate when integrated with the nuclear DNA. The efficiency of transformation of yeast cells by this type of vector is very low. However, this problem can be reduced by the use of hybrid *E. coli*—yeast vectors; the initial amplification of ligated vector and insert DNA sequences can be carried out in *E. coli*, subsequently transforming yeast cells with the plasmid DNA recovered from *E. coli.* Since high yields of plasmid DNA can readily be obtained from *E. coli* the low yield of yeast transformants per microgram of plasmid DNA is not a problem (see Section IV.D).

To recover the vector DNA (and cloned insert) from yeast transformants it is necessary to prepare high molecular weight yeast DNA and to digest this with a restriction endonuclease which is known to cut the plasmid DNA only once within or at the boundary of the yeast DNA insert. The mixture of DNA fragments is ligated at low DNA concentration, and the recircularized plasmid is recovered by transformation of *E. coli* cells, selecting for a marker on the bacterial component of the vector (Fig. 2). Depending upon the location of the recombination event which originally integrated the plasmid with respect to the yeast marker gene (*HIS* 3 in Fig. 2) and the restriction endonuclease recognition site on the vector DNA sequence (*Sal* I in Fig. 2), the mutant allele (originally located on the yeast genome) or the functional allele (from the vector) of the yeast marker gene will be recovered.

Although yeast transformation by integration vectors is relatively inefficient and the recovery of the integrated vector sequence from yeast transformants is tedious, there are some useful applications of integration vectors.

1. If the DNA from transformed yeast cells is digested with a restriction endonuclease which does not cut the vector DNA sequence, yeast nuclear DNA sequences flanking the site of integration can be readily recovered in *E. coli* along with the vector DNA. This procedure can be used for detailed analyses of specific regions of the yeast genome.

2. Fully characterized mutations can be introduced into a cloned gene using *in vitro* techniques and the mutant allele recombined into a yeast strain, replacing the original allele. Scherer and Davis (1979) used digestion with a DNA restriction endonuclease to introduce a

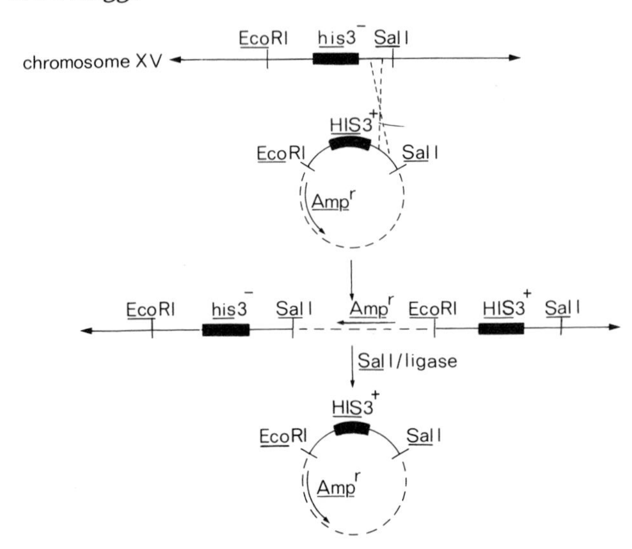

Figure 2 **Integration of a YIp vector into the genome of a yeast transformant, and recovery of the YIp DNA. A diagrammatic representation of a recombination event integrating the YIp5 plasmid DNA (Struhl *et al.*, 1979) at the *HIS* 3 locus on chromosome XV of yeast. The YIp5 DNA is recovered by digestion of the DNA and ligation as described in the text. N.B. If the DNA of this particular transformant was digested with *Eco* RI instead of *Sal* I, plasmids carrying the *his* 3^- allele would be recovered. —— yeast DNA; ▬ *HIS* 3 gene; — — — bacterial DNA sequence; Amp^r ampicillin resistance determinant.**

deletion mutation into a cloned *HIS* 3 gene. A *ura* 3^- yeast strain was transformed with the mutated gene (*his* 3∇) cloned on the non-replicating yeast vector YIp5 (Fig. 3). The fragment of yeast DNA containing the *URA* 3 marker gene on YIp5 provided the selection for transformants of a *ura* 3^- yeast strain. Being only about 1000 base pairs in length, there was very limited homology to direct integration at the *URA* 3 locus of the recipient (Scherer and Davis, 1979). This homology was further reduced by using a recipient strain of yeast with a deletion in the *URA* 3 gene. The YIp5—*his* 3∇ plasmid integrated preferentially at the *HIS* 3 locus in the recipient *ura* 3^- strain (Fig. 3). Transformants were selected as phenotypically Ura^+. After propagation of transformants for 10—20 generations on fully supplemented medium approximately 1% of the cells had lost one copy of the duplicated *his* 3 allele and all of the YIp5 vector sequence as a result of a recombination event essentially reversing the integration process. These recombinants were therefore *ura*⁻ as a result of losing the *URA* 3 marker on the vector sequence. A *ura*⁻ selection procedure (Bach and Lacroute, 1972) could be used to select these recombinants. The *ura*⁻ recombinants had an equal chance of retaining either the mutant or

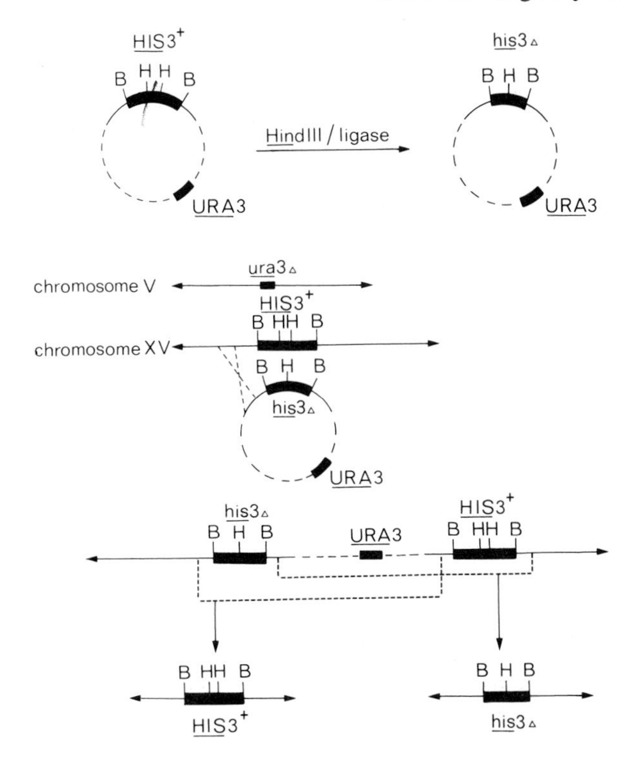

Figure 3 Outline procedure for replacement of *HIS* 3 in a yeast genome by a mutant *his* 3 allele constructed *in vitro*. The recombination event integrating the *his* 3∇ plasmid occurs preferentially at the *HIS* 3 locus on chromosome XV of the *ura* 3⁻ yeast recipient, as described in the text. During growth of the transformant in nonselective conditions two alternative recombination events can occur, eliminating one of the *his* 3 alleles and all of the vector DNA. ——— yeast DNA; ▬▬ *HIS* 3 or *URA* 3 genes; — — — bacterial DNA sequence; B, *Bam* H1 cleavage sites; H, *Hind* III cleavage sites. Based on Scherer and Davis (1979).

the functional *his* 3 allele. By this method a strain can be constructed carrying a fully characterized mutation in any gene (provided that gene can be cloned) and being otherwise isogenic to the parent strain, since no vector DNA is retained. Thus the phenotype of a characterized mutation can be determined (the reverse of classical genetic procedures), and stable deletion mutants can be created where previously no deletion mutations have been identified by classical genetic methods.

3. Foreign DNA sequences integrated into yeast chromosomal DNA or yeast genes translocated to new loci in the yeast genome can be used as biochemical or genetic markers to facilitate recombination studies of regions of the yeast genome not readily analysed

by conventional genetic techniques. This method has been used to study sister chromatid exchange in the ribosomal DNA of *S. cerevisiae* (Petes, 1980a; Szostak and Wu, 1980) which contains a cluster of approximately 140 copies of the repeated rDNA unit on chromosome XII. A *LEU* 2 gene was integrated by transformation in the rDNA cluster of a *leu* 2^- strain of yeast. Genetic and physical techniques were used to analyse recombination events between sister strands of rDNA gene clusters by measuring the deletion or duplication of the inserted *LEU* 2 marker. Unequal sister chromatid exchanges were observed at mitosis (Szostak and Wu, 1980) and in meiosis (Petes, 1980a) at a frequency sufficient to maintain the sequence homogeneity of the rDNA repeat units.

4. Amongst the bacterial plasmids carrying cloned yeast genes which had been isolated by complementation of *E. coli* mutants were some, e.g. YRp7 (*TRP* 1; Struhl *et al.*, 1979) and pYe (*arg* 4) (*ARG* 4; Hsiao and Carbon, 1979), which gave efficiencies of yeast transformation approximately 10^2—10^3-fold higher than usual. The yeast DNA sequences in these plasmids are believed to contain regions which function as origins of DNA replication (autonomously replicating sequences, *ars*) in *S. cerevisiae*. Although no evidence has been presented to show that these act as origins of replication in their normal environment in yeast, it has been estimated from such cloning studies (Beach *et al.*, 1980; Chan and Tye, 1980) that an *ars* sequence occurs on average once every 32—40 kb of DNA or 25 times per chromosome. These figures are in close agreement with an average yeast chromosomal DNA replicon size of 36 kb as determined by electron microscopy (Newlon and Burke, 1980). DNA sequences have been isolated from other eukaryotes, e.g. *Neurospora crassa*, *Dictyostelium discoideum*, *Caenhorabditis elegans*, *Drosophila melanogaster* and *Zea mays* which donated the property of autonomous replication in yeast cells to the YIp vector in which they were cloned (Stinchcomb *et al.*, 1980). It is not yet known whether these sequences promote DNA replication in the organism from which they were derived.

B *ars Vectors*

Plasmids containing *ars* sequences can be used as gene cloning vectors in *E. coli* and in *S. cerevisiae*. There are unique recognition sites in YRp7 (Fig. 4) for the DNA restriction endonucleases *Bam* HI and *Sal* I where additional DNA fragments can be inserted, resulting in inactivation of the tetracycline resistance gene. Recombinant plasmids can be selected in *E. coli* cells which become resistant to ampicillin or in *trp* 1^- strains of *S. cerevisiae* by growth on

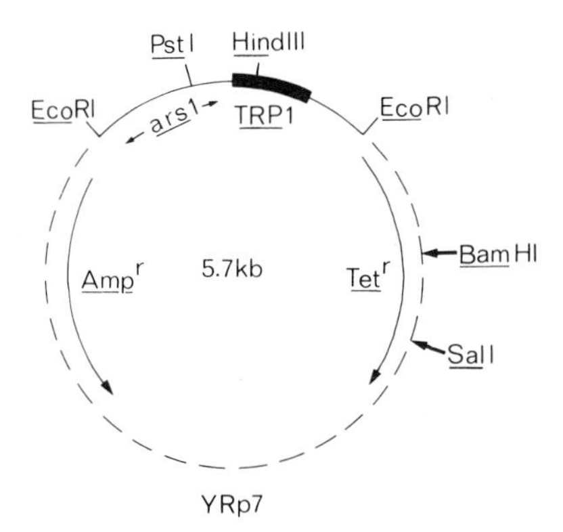

Figure 4 Physical map of *ars* vector YRp7. The diagram indicates the location of restriction endonuclease cleavage sites relative to the ampicillin resistance (Amp^r), tetracycline resistance (Tet^r) and *TRP* 1 genes, and the sequence directing autonomous replication (*ars* 1). ——— yeast DNA; ▬▬ *TRP* 1 gene; – – – pBR322 DNA sequence. Data from Stinchcomb *et al.* (1979).

tryptophan-deficient medium. Yeast transformants contain only one or a few copies of *ars* plasmid per cell but the covalently closed circular DNA in extracts of transformed clones can be readily recovered in large quantity after uptake and propagation in *E. coli* cells. *ars* Plasmids segregate very unstably at mitotic or meiotic division of yeast transformants. After propagation on tryptophan-deficient medium only 30—50% of the cells in transformed clones are still Trp^+ (Kingsman *et al.*, 1979). After propagation in tryptophan-supplemented medium only 10—20% of TRP^+ transformed cells retain the Trp^+ phenotype. When diploid cells carrying YRp7 are sporulated the germinated spores do not retain the replicon and all products of the meiotic segregation are trp^- (Kingsman *et al.*, 1979).

This feature of instability of *ars* plasmids was exploited by Clarke and Carbon (1980b) to identify a clone of yeast nuclear DNA including the centromeric region of chromosome III. Fragments of DNA from the region of the centromere-linked *LEU* 2, *CDC* 10 and *PGK* loci on chromosome III of *S. cerevisiae* were isolated from a pool of yeast—*E. coli* plasmid clones representing the entire yeast genome, by complementation of auxotrophic mutations in *E. coli* (*LEU* 2; Ratzkin and Carbon, 1977), complementation of mutations in yeast (*CDC* 10; Clark and Carbon, 1980a), immunological screening (*PGK*; Hitzemann *et al.*, 1979), and overlap hybridization (regions between these genes; Chinault and Carbon, 1979). Since *LEU* 2 maps to the left of the centromere of chromosome III and *PGK* maps to the right, a pool of cloned fragments extending over the region between these genes was assumed to include the centromeric region (*CEN* 3). A predicted property of a centromeric DNA sequence is that it should contribute stability to co-linear DNA during mitotic or meiotic division of the chromosomes. The cloned DNA fragment containing *CEN* 3 was identified by its ability to stabilize (*in cis*) plasmids also carrying a yeast chromosomal replicator (*ars* 1 or *ars* 2) making these behave like mini-chromosomes and segregate with Mendelian characteristics. The combination of a yeast replicator sequence (*ars*), a centromere and a selectable gene (i.e. a mini-chromosome) may prove to be a most useful and reliable gene cloning vector, especially where low copy number and high genetic stability are required.

C 2-Micron plasmid vectors

1 Properties of 2-micron DNA

The 2-micron plasmid, or *Scp* (Cameron *et al.*, 1977), is normally present in most if not all strains of *S. cerevisiae* and has been widely used for gene cloning in yeast. The properties of 2-micron DNA have been reviewed recently (Guerineau, 1979) and will only be outlined here. The majority of small circular DNAs in *S. cerevisiae* are approximately 2 μm in contour length as measured by electron microscopy although larger molecules, which are presumably dimers, trimers and recombination products of these, have also been observed (Stevens and Moustacchi, 1971; Guerineau *et al.*, 1971). There are up to 100 copies per cell of the plasmid DNA which has the same buoyant density as yeast nuclear DNA and can be purified in the supercoiled circular state by equilibrium centrifugation of yeast cell extract with caesium chloride and ethidium bromide (Guerineau, 1979).

In crosses between haploid strains with physically distinguishable species of 2-micron plasmid the 2-micron DNA from each haploid parent was found in all four meiotic progeny after sporulation of the diploids (Livingston, 1977). Thus, 2-micron DNA has a $4^+:0^-$ pattern of segregation typical of cytoplasmic genetic determinants. However, these genetic data do not provide information about the intracellular location of 2-micron DNA; rather they show that 2-micron DNA segregates in a non-Mendelian manner at meiotic division. Indeed, the intracellular location of the plasmid has been the subject of some controversy (Stevens and Moustacchi, 1971; Clark-Walker, 1972; Clark-Walker and Miklos, 1974). Livingston (1977) carried out crosses between strains with distinguishable nuclei and mitochondria, in which one parent carried a *kar* mutation, preventing fusion of the parental nuclei (Conde and Fink,

1976) and only one parent contained 2-micron DNA. Fifty per cent of the progeny carried 2-micron plasmid, which apparently segregated independently of the parental nucleus and mitochondria, suggesting a non-nuclear, non-mitochondrial location. However, in similar crosses using parents with distinguishable 2-micron plasmids Kielland-Brand *et al.* (1980) observed linked segregation of a 2-micron DNA species with the nucleus of parental origin. The possibility should be considered that 2-micron DNA moves between the nucleus and the cytoplasm, perhaps entering the nucleus to be replicated and transcribed.

Livingston and Hahne (1979) and Nelson and Fangman (1979) obtained evidence for 2-micron DNA molecules being organized in a chromatin structure with the same nucleosome repeat length as total yeast chromatin. These data and those of DNA replication studies described below are suggestive of a nuclear association of 2-micron DNA. The first evidence that 2-micron DNA replicated as small circular molecules came from the electron microscopy studies of Petes and Williamson (1975). In a preparation of yeast DNA with nuclear buoyant density they observed open double-branched circular DNA molecules (θ structures) with circumferences of approximately $2\,\mu m$. The branched structures were not observed in DNA preparations from yeast cultures which had been treated with an inhibitor of DNA replication. The effects on 2-micron DNA replication of some temperature sensitive mutants in the yeast cell division cycle (*cdc*) have been studied. *cdc* 8 and *cdc* 21 mutations which control nuclear and mitochondrial DNA replication also affect 2-micron DNA replication (Petes and Williamson, 1975). Nuclear and 2-micron DNA replication but not mitochondrial DNA replication were inhibited at the restrictive temperature in *cdc* 4, *cdc* 7 and *cdc* 28 mutants (Livingston and Kupfer, 1977). In addition, the yeast mating hormone, α factor, and cycloheximide, which prevent initiation of nuclear but not mitochondrial DNA synthesis, also inhibit 2-micron DNA synthesis (Livingston and Kupfer, 1977; Zeman and Lusena, 1974). Zakian *et al.* (1979) have observed that every 2-micron DNA molecule replicates once per cell division cycle and replication is restricted to early S phase. These data taken together indicate that 2-micron DNA replicates under cell cycle control, regulated by factors which also influence nuclear DNA replication.

The DNA sequencing data of Hartley and Donelson (1980) give the size of a cloned molecule of 2-micron plasmid as 6318 base pairs, although some strains of *S. cerevisiae* have been shown to contain plasmids slightly smaller than this as a result of small deletions in what appear to be otherwise similar sequences (Cameron *et al.*, 1977; Livingston, 1977). The plasmid DNA sequence contains a non-tandem inverted repeat of a sequence 598 base pairs long. The presence of the inverted repetition was clearly demonstrated by electron microscopy of molecular structures formed when single stranded plasmid DNA was allowed to form intramolecular base pairing. The inverted repeats self-annealed to form a double stranded stem flanked by the single stranded loops of the unique sequences (2347 and 2775 base pairs), resulting in a dumbell-like shape (Guerineau *et al.*, 1976; Hollenberg *et al.*, 1976; Livingston and Klein, 1977). Digestion of 2-micron DNA with the restriction endonuclease *Eco* RI results in the production of four DNA fragments (2246, 2407, 3911, 4072 base pairs; sizes from Hartley and Donelson, 1980) which are separable by agarose gel electrophoresis (Fig. 5). The sum of these fragment sizes is twice the size of the 2-micron DNA molecule, and this anomaly is a consequence of the co-existence of two forms of the plasmid molecule (A and B in Fig. 5) in yeast cells. The two forms of plasmid differ in the respective orientations of the unique sequences (i.e. the "loops" of the dumbell) although the sequences of the plasmids are otherwise the same (Livingston and Klein, 1977). Each unique segment of the plasmid DNA contains an *Eco* RI recognition sequence and therefore each of the two forms of the plasmid molecule has two *Eco* RI recognition sites in different relative positions, resulting in the cleavage of the DNA into two

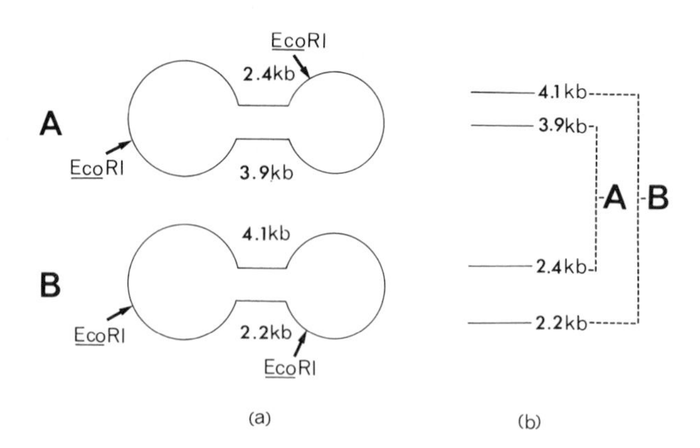

Figure 5 Physical maps of the two forms of 2-micron DNA with *Eco* RI cleavage sites. (a) the two forms (A and B) of 2-micron DNA in the dumbell representation to indicate the positions of the inverted repeat sequences (stem of dumbell). The large and small unique sequences are represented by the loops on the left and right respectively. The lengths of the *Eco* RI digestion fragments are indicated. (b) A diagrammatic representation of the banding pattern of fragments of *Eco* RI-digested 2-micron DNA after electrophoresis through agarose. The pairs of DNA fragments derived from form A or form B plasmid molecules are indicated.

fragments (2407 and 3911 base pairs from form A; 2246 and 4072 base pairs from form B; Fig. 5). The two types of plasmid normally co-exist in approximately equal proportions in yeast cells and are interconvertible as a result of intramolecular recombination (Beggs, 1978), which presumably occurs through the homology of the inverted repeat sequences. Molecular cloning of the 2-micron plasmid DNA in *E. coli* permitted the physical separation of the two forms of molecule and the construction of detailed maps of restriction endonuclease recognition sites (Beggs *et al.*, 1976; Hollenberg *et al.*, 1976; Gubbins *et al.*, 1977).

Plasmid-specific transcripts have been detected in 2-micron-containing ([2 μ]) yeast strains (Gubbins *et al.*, 1977; Guerineau, 1977; Broach *et al.*, 1979a). Broach *et al.* (1979a) mapped poly A-containing 2-micron specific transcripts and showed that both strands of 2-micron DNA are extensively transcribed *in vivo.* The 2-micron specific transcripts isolated from yeast cells, when translated *in vitro* directed the synthesis of polypeptides of apparent molecular weights 30 000, 33 000, 37 000 and 60 000 daltons. However, 2-micron protein products have not yet been identified *in vivo*.

From DNA sequence data, Hartley and Donelson (1980) identified potential protein coding regions. These are lengths of sequence starting with an ATG translation initiation codon, followed by sufficient sense codons to code for a polypeptide before the occurrence of one of the three translation stop codons. Three long open reading frames were found and named Able, Baker and Charlie (Fig. 6). These have potential coding capacity for proteins with molecular weights of 48 625, 43 235 and 33 199 daltons respectively. A possible function for the hypothetical protein Able has been proposed (Broach and Hicks, 1980; see below).

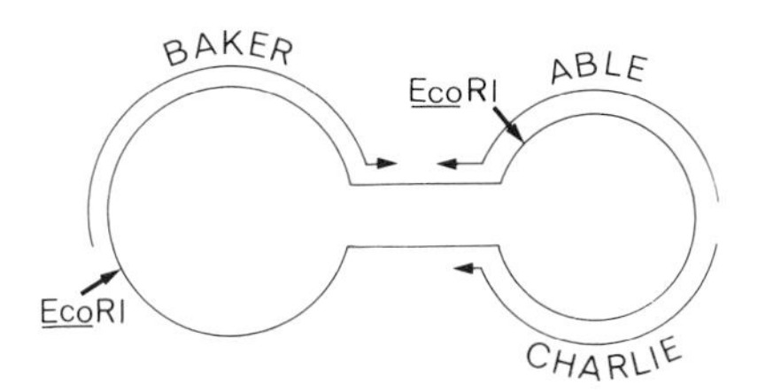

Figure 6 The locations of translation open reading frames in 2-micron DNA. Type A 2-micron plasmid DNA is shown in the dumbell representation, indicating the locations of the translation open reading frames Able, Baker and Charlie as determined by Hartley and Donelson (1980). Able is encoded by the DNA strand of opposite polarity to Charlie and Baker as indicated by the arrows.

2 *Transformation with recombinant 2-micron plasmids*

Hybrid plasmids containing the entire 2-micron DNA sequence, the yeast *LEU* 2 gene and a bacterial vector sequence, e.g. pJDB219 and pJDB248 (Fig. 7; Beggs, 1978) transform *leu* 2^- yeast strains with high frequency (10^3–10^5 transformants per microgram of DNA). Extraction of the DNA from yeast transformants, electrophoresis through agarose, transfer to nitrocellulose and hybridization with radioactively labelled plasmid DNA showed the presence of the hybrid plasmid DNAs in the supercoiled circular form, indicating that they were replicating in transformed yeast cells (Beggs,

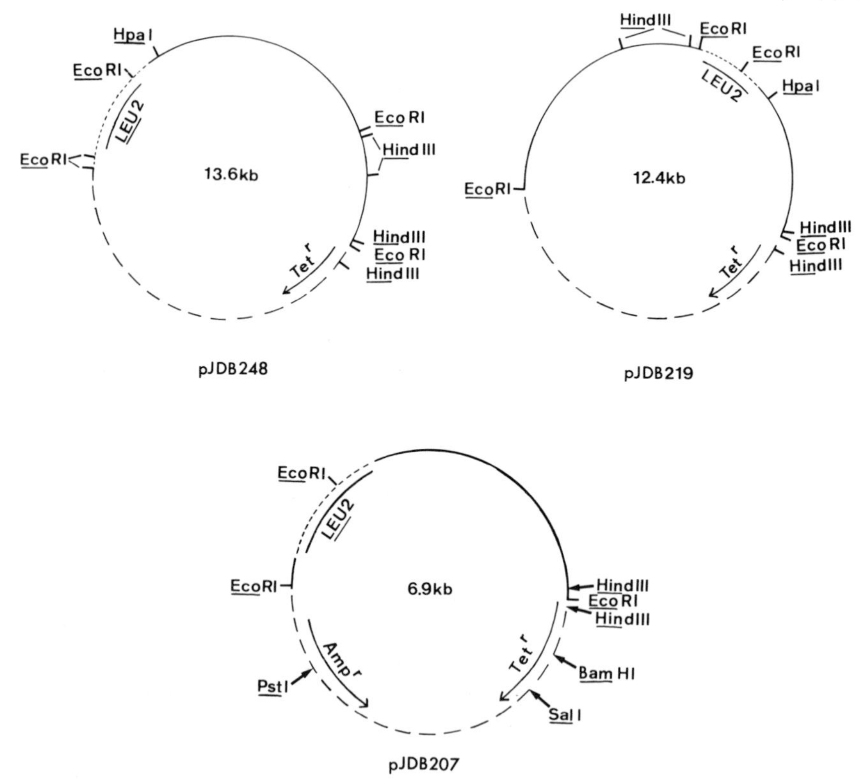

Figure 7 Physical maps of pJDB248, pJDB219 and pJDB207. pJDB248 (form A 2-micron sequence) and pJDB219 (form B 2-micron sequence) were constructed as described by Beggs (1978). pJDB207 is a derivative of pJDB219 (Beggs, 1981), constructed for use as a gene cloning vector. Locations of restriction endonuclease cleavage sites are shown relative to the *LEU* 2 gene and ampicillin (*Amp*^{r}) and tetracycline (*Tet*^{r}) resistance determinants. ——— 2-micron DNA sequence; -------- yeast DNA sequence; — — — bacterial DNA sequence (pMB9 in pJDB219 and pJDB248; pAT153 (Twigg and Sherratt, 1980) in pJDB207).

1978). The efficiency of yeast transformation with pJDB248 is greater than for pJDB219, and pJDB248-transformed clones grow faster on selective medium. However, the copy number of pJDB219 in yeast cells is higher than for pJDB248, and pJDB219-transformed clones are genetically more stable (Beggs, 1978 and unpublished results). It is not understood why these two plasmids have such different properties in yeast transformation. One possibility is that the *LEU* 2 gene on pJDB219 is poorly expressed, causing slower growth of transformants on selective medium and causing the plasmid copy number and *LEU* 2 gene dosage to increase with a resultant increase in the efficiency of plasmid segregation at cell division.

pJDB219 contains the 2-micron sequence in form B (Fig. 5) ligated to pMB9 through the *Eco* RI site of the small unique region. [$2\mu^+$] yeast transformants contain both the A and B forms of the hybrid plasmid, confirming the hypothesis that A and B forms of 2-micron DNA result from a recombinational event interconverting the two forms in the yeast cell (Beggs, 1978). However this interconversion of pJDB219 does not occur in [$2\mu^0$] transformants (Dobson *et al.*, 1980a). Blanc *et al.* (1979) and Broach and Hicks (1980) showed that this lack of interconversion of the plasmid forms in transformed [$2\mu^0$] clones is a property confined to hybrid plasmids in which the 2-micron sequence is interrupted at the *Eco* RI recognition site in the small unique region (as in pJDB219). Broach and Hicks (1980) suggest that this *Eco* RI site lies within a gene *FLP* encoding a protein required for intramolecular recombination of 2-micron DNA. In [$2\mu^+$] strains the *flp*$^-$ defect would be complemented *in trans.* This gene corresponds to the hypothetical gene Able of Hartley and Donelson (1980).

Various combinations of yeast—*E. coli* sequences containing part or all of the 2-micron DNA have been constructed and used in yeast transformation studies. A plasmid carrying any one of the four *Eco* RI fragments of 2-micron DNA (Fig. 5) transforms [$2\mu^+$] yeast strains with high frequency. However, the stability of the transformed phenotype varies depending upon which part of the 2-micron sequence is used (Hicks *et al.*, 1979a; Broach and Hicks, 1980). Plasmids containing the 2.2 kb or 3.9 kb *Eco* RI fragments of 2-micron DNA are relatively stably maintained in transformed clones, whereas the 2.4 kb and 4.1 kb *Eco* RI fragments give plasmids with a highly unstable phenotype. Some 2-micron-free ([$2\mu^0$]) strains of *S. cerevisiae* have been derived by loss of 2-micron DNA from [$2\mu^+$] strains (Guerineau *et al.*, 1974; Gerbaud and Guerineau, 1980; Dobson *et al.*, 1980b) and these as well as [$2\mu^0$] *S. carlsbergensis* strains (Livingston, 1977) have been tested as recipient strains for transformation by 2-micron vectors. Plasmids containing the entire 2-micron sequence transform these [$2\mu^0$] strains with high efficiency and the transformed phenotype is relatively stable (Blanc *et al.*, 1979; Broach and Hicks, 1980; Dobson *et al.*, 1980a). The plasmids with the 2.2 kb or 3.9 kb *Eco* RI fragments of 2-micron DNA are also efficient in transformation; those containing the 2.4 kb and 4.1 kb *Eco* RI fragments fail to yield transformants.

Using *in vitro* generated deletion mutants, Broach and Hicks (1980) identified a 2-micron DNA sequence approximately 350 base pairs long extending from the larger "unique sequence" (the left-hand loop of the dumbells in Fig. 5) into the inverted repeat sequence of the 2.2 kb or 3.9 kb *Eco* RI fragments which appeared to act as an origin of DNA replication in yeast and in an *in vitro* DNA replication assay (Broach, 1981). This region of the plasmid includes a 122 base pair sequence in the inverted repeat containing symmetry elements which can be drawn as large secondary structures with stem and loop configurations typical of those found near the origins of replication of many genomes (Hartley and Donelson, 1980).

The relatively high frequency of transformation of [$2\mu^+$] strains by plasmids with the 2.4 kb and 4.1 kb *Eco* RI fragments of 2-micron DNA apparently results from the homology between these plasmids and the resident 2-micron plasmid which allows recombination to occur. The resultant chimaeric plasmid, containing the resident 2-micron sequence and the transforming vector, replicates and can subsequently dissociate into two different plasmids again, thus giving the impression that the vector plasmid replicated and also facilitating the selection of transformed clones.

Various combinations of yeast or *E. coli*—yeast hybrid sequences including part or all of the 2-micron sequence have therefore been used in yeast transformation but only a few of these plasmids are convenient for use as cloning vectors. The smaller and therefore more versatile plasmid vectors (e.g. YEp13, Broach *et al.*, 1979b; pFL1-4, Chevallier *et al.*, 1980; pJDB207, Beggs, 1981, Fig. 7) contain only the 2.2 kb *Eco* RI fragment or a similar fragment of 2-micron DNA to provide a high frequency of yeast transformation and multiple copies of replicating plasmid DNA in yeast cells. These plasmids have mostly been used to transform yeast strains already containing a 2-micron plasmid and one complication which arises in the characterization of the DNA of yeast transformants is the recombination which occurs readily between the resident 2-micron plasmid and the incoming 2-micron vector, resulting in the formation of complex recombinant plasmid molecules. If the DNA sequence under study (gene X) and the selectable marker gene of the vector are not contiguous in the cloned plasmid there is a significant probability that they will segregate into separate recombinant plasmids in the yeast cells. In these circumstances maintenance of selection for the vector marker gene does not ensure the maintenance of gene X. It should, therefore, be borne in mind that testing transformant clones for the phenotype contributed by the marker gene does not, by itself, provide reliable evidence of the stability of the cloning vector and does not distinguish between the presence of the vector marker gene in the autonomously replicating state as opposed to integrated in the yeast nuclear chromosomal DNA.

The use of [$2\mu^0$] yeast strains as recipients for DNA transformation may reduce the recombination problems. However, at present, few data are available on the behaviour of 2-micron vectors in [$2\mu^0$] recipient strains. It appears that most 2-micron vectors which are reasonably stably maintained in [$2\mu^+$] recipient strains are less stable in [$2\mu^0$] strains and are rapidly lost even when the clones are cultured under conditions selecting for plasmid maintenance. This instability in [$2\mu^0$] strains is probably the result of inefficient replication and/or segregation of the vector DNA and may be explained by the absence of part of the 2-micron sequence as proposed by Broach and Hicks (1980). Even so, 2-micron vectors in [$2\mu^+$] or [$2\mu^0$] strains tend to be more stably maintained than *ars* vectors without a centromere, and have the useful property of multiple copy number in yeast which may be especially important where a high level of expression of a cloned gene is required.

Beach and Nurse (1981) have successfully used 2-micron vectors to transform *S. pombe*. pJDB248 which transforms *leu* 2^- strains of *S. cerevisiae* with high frequency also efficiently transforms and replicates in *leu* 1^- strains of *S. pombe* (presumably the *LEU* 1 gene of *S. pombe* codes for β-isopropylmalate dehydrogenase). However, other 2-micron vectors such as pJDB219 and its derivative pJDB207 which efficiently transform *S. cerevisiae* and replicate with high copy number (Beggs, unpublished results) do not transform *S. pombe* efficiently. A derivative of pJDB248, pDB248, has been constructed by Beach and Nurse (1981) for use as a cloning vector with *S. pombe*.

D The use of hybrid yeast—*E. coli* vectors

The vectors described above are all chimaeras of yeast and *E. coli* DNA sequences. The initial construction of these vectors was facilitated by the use of *E. coli* as the host in which the DNA was cloned and propagated especially as several of these were made before the method for yeast transformation was developed. It is now possible to carry out all stages of gene cloning directly in yeast.

Kielland-Brand *et al.* (1979) cloned the yeast *HIS* 4 gene by ligating yeast DNA fragments to linearized 2-micron DNA and directly transforming a *his* 4 yeast mutant. A single HIS^{+} transformant was obtained using 24 μg of ligated yeast DNA. Although this yield could probably be improved by optimizing the experimental conditions, this direct approach is generally impractical with current procedures for the reasons of economy described below, and is likely to be used only where the application of hybrid bacterial—yeast vectors is considered undesirable.

Depending upon which type of yeast cloning system is used, purified plasmid DNAs will yield between 1 (at least, with integrative vectors) and 10^5 (the maximum yield so far for 2-micron vectors) yeast transformants per microgram of supercoiled plasmid DNA. Using *E. coli* strains one can expect to obtain from 10^5 to 10^8 transformants per microgram of supercoiled plasmid DNA, depending mainly upon the recipient strains and the method used for the preparation of competent cells. (Yields of transformants are generally reduced to between 1% and 50% of these values when ligated DNA fragments are used.) Since, for reasons of economy of materials, labour and time one prefers to have the highest possible yield of transformants per microgram of DNA to be cloned (often the DNA is the most precious component; however, enzymes used in the *in vitro* fractionation and recombination reactions are also expensive) it is clearly more efficient to first amplify recombinant plasmids in *E. coli.* Each *E. coli* clone can subsequently be propagated either individually or in a pool of mixed clones and large quantities of purified recombinant plasmid DNAs can be easily and economically prepared. The purified DNA can subsequently be used to obtain the desired number of yeast transformants. Plasmids of interest (e.g. containing a gene selected by complementation in yeast) can be isolated from yeast cell extracts. However, the yield of plasmid DNA from yeast cultures depends both on the type of yeast vector and on the particular yeast strain used and is very low relative to plasmid DNA yields obtained routinely and more rapidly from *E. coli.* It is therefore sometimes practical to transfer plasmid DNAs from yeast to *E. coli* (by transformation of *E. coli* with relatively crude yeast cell extracts) for the preparation of large quantities of a specific plasmid DNA.

Broach *et al.* (1979b) and Kingsman *et al.* (1979) have simplified the transfer of DNA from *E. coli* to yeast, mixing partially lysed *E. coli* protoplasts with yeast sphaeroplasts (transfusion). However, the reported yield of yeast transformants is relatively low. Although discussion of hybrid yeast—*E. coli* vectors has so far been limited to the use of *E. coli* plasmids as the bacterial moiety, hybrid yeast–bacteriophage lambda (Davis, 1981) and yeast–cosmid vectors (Hohn and Hinnen, 1980) have also been used.

It is therefore now feasible to clone any yeast gene by transformation into yeast using a pool of plasmid clones constructed with one of the types of vectors described above. The efficiencies of yeast transformation obtained with different types of vector are summarized in Table 2. The size of the collection of clones required to ensure the inclusion of every yeast DNA sequence depends on the average length of DNA sequence cloned and may be calculated as described by Clarke and Carbon (1976).

V Yeast transformation without cloning vectors

The uptake and stable maintenance of high molecular weight yeast DNA by recipient yeast strains appears to be possible without the use of *in vitro* genetic manipulation of plasmids. Barney *et al.* (1980a) demonstrated the transfer of the ability to utilize dextrin to non-dextrin-utilizing brewing strains of *S. cerevisiae* and *S. uvarum* via partially purified high molecular weight DNA from a dextrin-utilizing non-brewing yeast *S. diastaticus.* They also introduced the property of flocculence into a non-flocculent yeast by transformation with DNA isolated from a flocculent yeast (Barney *et al.*, 1980b). Since flocculent transformants could not be selected on solid medium and it would have been impractical to screen for flocculence in liquid cultures, a system of co-transformation was used. *ADE* 1 (a gene in the adenine biosynthetic pathway) maps close to (approximately 40 cM from) *FLO* 1 (a flocculence determinant). An *ade* 1 *flo* recipient strain was transformed with DNA from an *ADE* 1 *FLO* 1 donor and Ade^+ transformants were selected. A yield of 105 Ade^+ transformants was obtained per 50 μg of donor DNA. Ade^+ transformants were screened for flocculence and 8% were found to have been co-transformed with the *ADE* 1 and *FLO* 1 genes. Tetrad analysis of the co-transformants showed Mendelian segregation of the newly acquired genes, indicating stable integration into the nuclear DNA of the recipient cells. (It has also been observed that when yeast cells are transformed with a mixture of two types of plasmid DNAs many transformants are co-transformed by both plasmid DNAs (Hicks *et al.*, 1979; Jimenez and Davies, 1980).)

VI Applications

Methods for cloning genes by transformation directly into yeast have facilitated the isolation of several yeast genes which could not

Table 2 Efficiencies of yeast transformation with different types of vector. The nomenclature YIp (yeast integrating plasmid), YRp (yeast replicon plasmid) and YEp (yeast episomal plasmid) is that of Struhl *et al.* (1979).

Type of vector				Yeast transformation: Frequency (proportion of viable sphaeroplasts transformed)	Yeast transformation: Efficiency (transformants per μg DNA)
(YIp)	bacterial vector	+	selectable yeast gene (e.g. *LEU* 2, *HIS* 3 or *URA* 3)	10^{-7}	1–10
(YRp)	bacterial vector	+	yeast origin of DNA replication and selectable yeast gene (e.g. *Eco* RI fragment including *TRP* 1)	10^{-4}	10^{2}–10^{4}
(YEp)	bacterial vector	+	selectable gene + "2-micron DNA"	10^{-3}	10^{3}–10^{5}

have been readily isolated by cloning in *E. coli.* The yeast *HIS* 4 gene is complex, coding for a multifunctional protein providing three enzyme activities in the histidine biosynthetic pathway. The *HIS* 4 gene was cloned by complementation of a *his* 4 mutant strain of yeast (Hinnen *et al.*, 1979) after exhaustive attempts to isolate the yeast gene by complementation of *his* mutants of *E. coli* had failed.

Many genes affecting the yeast cell division cycle have been mapped through the isolation of conditional lethal mutations. However, very few of these *CDC* genes have been characterized in terms of protein products and specific functions. By complementation of *cdc* mutants in non-permissive growth conditions cell cycle genes can be cloned in yeast (Nasmyth and Reed, 1980; Clarke and Carbon, 1980a). Hopefully, the availability of cloned *CDC* genes will facilitate the identification of the gene products and their functions and help to elucidate the mechanism by which the yeast cell division cycle is regulated.

One of the most fascinating control mechanisms operating in yeast is that of mating type switching, the interconversion of yeast cells from one mating type (*a* or α) to the other (Mortimer and Hawthorne, 1969). This is of particular interest as it may provide an important model for a mechanism of cellular differentiation. The development of yeast transformation techniques has led to the cloning of a set of genes involved in mating type interconvertion and the biochemical confirmation of the "cassette model" for mating type switching which had been proposed on the basis of complex genetic analyses (Hicks *et al.*, 1977, 1979b; Nasmyth and Tatchell, 1980). In this model the genes controlling yeast cell mating type reside on transposable elements which are only expressed when transposed from silent storage sites to the mating type locus some distance away on the same chromosome. It has become apparent from DNA sequencing studies of the cloned mating type genes that a simple model of control of transcription through closely linked promoter sites cannot explain the differential regulation of transcription of identical genes occupying different locations on the genome, and position effect models of gene expression are now being investigated (Klar *et al.*, 1981; Nasmyth *et al.*, 1981).

Many other yeast genes have now been cloned and the applications of yeast cloning techniques have been richly rewarded by the increased knowledge of yeast molecular biology (for reviews see Petes, 1980b; Strathern *et al.*, 1981; Davis *et al.*, 1981). Also, these techniques which permit short segments of DNA to be introduced into any yeast strain are likely to be heavily exploited in the brewing industry. It would be particularly valuable to be able to introduce a specific property into an existing brewing strain without affecting other brewing properties of the strain which may be difficult to protect in classical hybridization procedures, and whose phenotypes may be difficult to identify in progeny except by time consuming and tedious assays.

It is still uncertain to what extent yeast cells will be useful recipients in which to clone genes from other eukaryotes and obtain functional expression. Henikoff *et al.* (1981) cloned a *Drosophila* gene in yeast by complementation of an adenine-8 mutation. Transcription of the *Drosophila* gene in yeast was apparently initiated within the *Drosophila* DNA sequence and transformants carrying the cloned *Drosophila* gene grew on adenine deficient medium at about two-thirds the rate of wild type yeast cells. Yeast cells transformed with a human interferon gene transcribed from a yeast promoter sequence produce functional interferon (B. D. Hall, personal communication). Thus, yeast cells can transcribe certain non-yeast genes of eukaryotic origin and at least in these two examples the yeast ribosomes were able to translate the messenger RNAs into functional protein.

Neither of these genes has an interrupted coding sequence and a question of great interest is whether yeast cells can efficiently splice precursor RNAs copied from non-yeast genes containing intervening sequences. The rabbit β-globin gene cloned on a 2-micron vector in yeast was transcribed, but transcription terminated prematurely within an intervening sequence and no spliced RNA was detected (Beggs *et al.*, 1980). At present there is insufficient information about the mechanism and degree of specificity of RNA splicing reactions to determine the reason for this failure. However, the yeast actin gene has recently been cloned and found to contain an intervening sequence. This is the first example of a yeast nuclear gene with an interrupted coding sequence (with the exception of several tRNA genes) and electron microscopy and DNA sequencing data show that the sequences at the intron boundaries resemble the prototype sequences found in genes of higher eukaryotes (Gallwitz and Sures, 1980). It seems likely, therefore, that the yeast splicing machinery will resemble that in other eukaryotes. It remains to be seen to what extent yeast splicing enzymes will function with "foreign" gene transcripts and extend the application of yeast cloning techniques to the isolation and expression of eukaryotic genes in general.

VII References

Bach, M. L. and Lacroute, F. (1972). Direct selective techniques for the isolation of pyrimidine auxotrophs in yeast. *Mol. Gen. Genet.* **115**, 126–130.

Bach, M. L., Lacroute, F. and Botstein, D. (1979). Evidence for transcriptional regulation of orotidine-5′-phosphate decarboxylase in yeast by hybridisation of mRNA to the yeast structural gene cloned in *Escherichia coli*. *Proc. Natn. Acad. Sci. U.S.A.* **76**, 386–390.

Barney, M., Jansen, G. P. and Helbert, J. R. (1980a). Use of spheroplast fusion and genetic transformation to introduce dextrin utilisation into *Saccharomyces uvarum*. *J. Am. Soc. Brewing Chem.* **38**, 1–5.

Barney, M., Jansen, G. P. and Helbert, J. R. (1980b). Use of genetic transformation for the introduction of flocculence into yeast. *J. Am. Soc. Brewing Chem.* **38**, 71–74.

Beach, D. and Nurse, P. (1981). High-frequency transformation of the fission yeast *Schizosaccharomyces pombe*. *Nature, Lond.* **290**, 140–142.

Beach, D., Piper, M. and Shall, S. (1980). Isolation of chromosomal origins of replication in yeast. *Nature, Lond.* **284**, 185–187.

Beggs, J. D. (1978). Transformation of yeast by a replicating hybrid plasmid *Nature, Lond.* **275**, 104–109.

Beggs, J. D. (1981). Multiple-copy yeast plasmid vectors. *In* "Molecular Genetics in Yeast", Alfred Benzon Symposium, Munksgaard (Eds D. von Wettstein, J. Friis, M. Kielland-Brand and A. Stenderup) Vol. 16. In press.

Beggs, J. D., Guerineau, M. and Atkins, J. F. (1976). A map of the restriction targets in yeast 2 micron plasmid DNA cloned in bacteriophage lambda. *Mol. Gen. Genet.* **148**, 287–294.

Beggs, J. D., Van Den Berg, J. Van Ooyen, A. and Weissmann, C. (1980). Abnormal expression of chromosomal rabbit β-globin gene in *Saccharomyces cerevisiae*. *Nature, Lond.* **283**, 835–840.

Blanc, H., Gerbaud, C., Slonimski, P. and Guerineau, M. (1979). *Mol. Gen. Genet.* **176**, 335–342.

Broach, J. R. (1981). Replication functions associated with the yeast plasmid 2μ circle. *In* "Molecular Genetics in Yeast", Alfred Benzon Symposium, Munksgaard (Eds D. von Wettstein, J. Friis, M. Kielland-Brand and A. Stenderup) Vol. 16. In press.

Broach, J. R. and Hicks, J. B. (1980). Replication and recombination functions associated with the yeast plasmid 2μ circle. *Cell* **21**, 501–508.

Broach, J. R., Atkins, J. F., McGill, C. and Chow, L. (1979a). Identification and mapping of the transcriptional and translational products of the yeast plasmid 2μ circle. *Cell* **16**, 827–839.

Broach, J. R., Strathern, J. N. and Hicks, J. B. (1979b). Transformation in yeast: development of a hybrid cloning vector and isolation of the *CAN* 1 gene. *Gene* **8**, 121–133.

Cameron, J. R., Philippsen, P. and Davis, R. W. (1977). Analysis of chromosomal integration and deletions of yeast plasmids. *Nucl. Acids Res.* **4**, 1429–1448.

Chan, C. S. M. and Tye, B-K. (1980). Autonomously replicating sequences in *Saccharomyces cerevisiae*. *Proc. Natn. Acad. Sci. U.S.A.* **77**, 6329–6333.

Chevallier, M-R., Bloch, J-C. and Lacroute, F. (1980). Transcriptional and translational expression of a chimeric bacterial–yeast plasmid in yeast. *Gene* **11**, 11–19.

Chinault, A. C. and Carbon, J. (1979). Overlap hybridisation screening: isolation and characterisation of overlapping DNA fragments surrounding the *leu* 2 gene on yeast chromosome III. *Gene* **5**, 111–126.

Clarke, L. and Carbon, J. (1976). A colony bank containing synthetic ColE1 hybrid plasmids representative of the entire *E. coli* genome. *Cell* **9**, 91–99.

Clarke, L. and Carbon, J. (1978). Functional expression of cloned yeast DNA in *Escherichia coli*: specific complementation of argininosuccinate lyase (*arg*H) mutations. *J. Mol. Biol.* **120**, 517–532.

Clarke, L. and Carbon, J. (1980a). Isolation of the centromere-linked *CDC* 10 gene by complementation in yeast. *Proc. Natn. Acad. Sci. U.S.A.* **77**, 2173–2177.

Clarke, L. and Carbon, J. (1980b). Isolation of a yeast centromere and construction of functional small circular chromosomes. *Nature, Lond.* **287**, 504–509.

Clark-Walker, G. D. (1972). Isolation of circular DNA from a mitochondrial fraction from yeast. *Proc. Natn. Acad. Sci. U.S.A.* **69**, 388–392.

Clark-Walker, G. D. and Miklos, G. L. (1974). Localisation and quantification of circular DNA in yeast. *Eur. J. Biochem.* **41**, 359–365.

Cohen, J. D., Eccleshall, T. R., Needleman, R. B., Federoff, H., Buchferer, B. A. and Marmur, J. (1980). Functional expression in yeast of the *Escherichia coli* plasmid gene coding for chloramphenicol acetyltransferase. *Proc. Natn. Acad. Sci. U.S.A.* **77**, 1078–1082.

Conde, J. and Fink, G. R. (1976). A mutant of *Saccharomyces cerevisiae* defective for nuclear fusion. *Proc. Natn. Acad. Sci. U.S.A.* **73**, 3651–3655.

Davis, R. W. (1981). DNA transformation of *Saccharomyces cerevisiae*. *In* "Molecular and Cellular Biology". ICN-UCLA 10th Symposium. In press.

Davis, R. W., Fink, G. R. and Botstein, D. (1981). *A. Rev. Biochem.* **50**, in press.

Dobson, M., Futcher, A. B. and Cox, B. S. (1980a). Control of recombination within and between DNA plasmids of *Saccharomyces cerevisiae*. *Curr. Genet.* **2**, 193–200.

Dobson, M. J., Futcher, A. B. and Cox, B. S. (1980b). Loss of 2μm DNA from *Saccharomyces cerevisiae* transformed with the chimaeric plasmid pJDB219. *Curr. Genet.* **2**, 201–204.

Gallwitz, D. and Sures, I. (1980). Structure of a split yeast gene: complete nucleotide sequence of the actin gene in *Saccharomyces cerevisiae*. *Proc. Natn. Acad. Sci. U.S.A.* **77**, 2546–2550.

Gerbaud, C. and Guerineau, M. (1980). 2μm plasmid copy number in different strains and repartition of endogenous and 2μm chimeric plasmids in transformed strains. *Curr. Genet.* **1**, 219–228.

Gubbins, E. J., Newlon, C. S., Kann, M. D. and Donelson, J. E. (1977). Sequence organisation and expression of a yeast plasmid DNA. *Gene* **1**, 185–207.

Guerineau, M. (1977). Expression of a yeast episome: RNA-DNA hybridisation studies. *FEBS Letters* **80**, 426–428.

Guerineau, M. (1979). Plasmid DNA in yeast. *In* "Viruses and Plasmids in Fungi", Series on Mycology, (Ed. P. A. Lemke) Vol. 1. Marcel Dekker, New York and Basel.

Guerineau, M., Grandchamp, C., Paoletti, C., Slonimski, P. P. (1971). Characterisation of a new class of circular DNA molecules in yeast. *Biochem. Biophys. Res. Commun.* **42**, 550–557.

Guerineau, M., Slonimski, P. P. and Avner, P. R. (1974). Yeast episome: oligomycin resistance associated with a small covalently closed non-mitochondrial circular DNA. *Biochem. Biophys. Res. Commun.* **61**, 462–469.

Guerineau, M., Grandchamp, C. and Slonimski, P. P. (1976). Circular DNA of a yeast episome with two inverted repeats: structural analysis by a restriction enzyme and electron microscopy. *Proc. Natn. Acad. Sci. U.S.A.* **73**, 3030–3034.

Hartley, J. L. and Donelson, J. E. (1980). Nucleotide sequence of the yeast plasmid. *Nature, Lond.* **286**, 860–865.

Hawthorne, D. C. and Mortimer, R. K. (1978). Genetic map of *Saccharomyces cerevisiae.* "Handbook of Biochemistry and Molecular Biology", Vol. 2, 765–832.

Henikoff, S., Tatchell, K., Hall, B. D. and Nasmyth, K. A. (1981). Isolation of a gene from *Drosophila* by complementation in yeast. *Nature, Lond.* **289**, 33–37.

Hicks, J. B., Strathern, J. and Herskowitz, T. (1977). The cassette model of mating type interconversion. *In* "DNA Insertion Elements, Plasmids and Episomes" (Eds A. Bukhari, J. Shapiro and S. Adhya) pp. 475–462. Cold Spring Harbor, New York.

Hicks, J. B., Hinnen, A. and Fink, G. R. (1979a). Properties of yeast transformation. *Cold Spring Harb. Symp. Quant. Biol.* **43**, 1305–1313.

Hicks, J. B., Strathern, J. N. and Klar, A. J. S. (1979b). Transposable mating type genes in *S. cerevisiae. Nature, Lond.* **282**, 478–482.

Hinnen, A., Hicks, J. B. and Fink, G. R. (1978). Transformation of yeast. *Proc. Natn. Acad. Sci. U.S.A.* **75**, 1929–1933.

Hinnen, A., Farabaugh, P., Ilgen, C. and Fink, G. R. (1979). ICN-UCLA Symp. 14, 43–50.

Hitzemann, R. A., Chinault, A. C., Kingsman, A. J. and Carbon, J. (1979). ICN-UCLA Symp. 14, 57–68.

Hohn, B. and Hinnen, A. (1980). Cloning with cosmids in *E. coli* and yeast. *In* "Genetic Engineering. Principles and Methods" (Eds J. K. Setlow and A. Hollaender) Vol. 2, pp. 169–183. Plenum Press, New York.

Hollenberg, C. P. (1979). The expression of bacterial antibiotic resistance genes in the yeast *Saccharomyces cerevisiae. In* "Plasmids of Medical, Environmental and Commercial Importance" (Eds K. N. Timmis and A. Puhler) pp. 481–492. Elsevier/North-Holland Biomedical Press, Amsterdam.

Hollenberg, C. P., Degelmann, A., Kustermann-Kuhn, B. and Royer, H. D. (1976). Characterisation of 2μm DNA of *Saccharomyces cerevisiae* by restriction fragment analysis and integration in an *Escherichia coli* plasmid. *Proc. Natn. Acad. Sci. U.S.A.* **73**, 2072–2076.

Hsiao, C-L. and Carbon, J. (1979). High-frequency transformation of yeast by plasmids containing the cloned yeast *ARG* 4 gene. *Proc. Natn. Acad. Sci. U.S.A.* **76**, 3829–3833.

Jimenez, A. and Davies, J. (1980). Expression of a transposable antibiotic resistance element in *Saccharomyces. Nature, Lond.* **287**, 869–871.

Kielland-Brand, M. C., Nilsson-Tillgren, T., Holmberg, S., Petersen, J. G. L. and Svenningsen, B. A. (1979). Transformation of yeast without the use of foreign DNA. *Carlsberg Res. Commun.* **44**, 77–87.

Kielland-Brand, M. C., Wilken, B., Holmberg, S., Petersen, J. G. L. and Nilsson-Tillgren, T. (1980). Genetic evidence for nuclear location of 2-micron DNA in yeast. *Carlsberg Res. Commun.* **45**, 119–124.

Kingsman, A. J., Clarke, L., Mortimer, R. K. and Carbon, J. (1979). Replication in *Saccharomyces cerevisiae* of plasmid pBR313 carrying DNA from the yeast *TRP* 1 region. *Gene* **7**, 141–153.

Klar, A. J. S., Strathern, J. N., Broach, J. R. and Hicks, J. B. (1981). Regulation of transcription in expressed and unexpressed mating type cassettes of yeast. *Nature, Lond.* **289**, 239–244.

Livingston, D. M. (1977). Inheritance of the 2μm DNA plasmid from *Saccharomyces. Genetics* **86**, 73–84.

Livingston, D. M. and Hahne, S. (1979). Isolation of a condensed intracellular form of the 2μm DNA plasmid of *S. cerevisiae. Proc. Natn. Acad. Sci. U.S.A.* **76**, 3727–3731.

Livingston, D. M. and Klein, H. L. (1977). Deoxyribonucleic acid sequence organisation of yeast plasmid. *J. Bacteriol.* **129**, 472–481.

Livingston, D. M. and Kupfer, D. M. (1977). 'Control of *Saccharomyces cerevisiae* 2μm DNA replication by cell division cycle genes that control nuclear DNA replication'. *J. Mol. Biol.* **116**, 249–260.

Mortimer, R. K. and Hawthorne, D. C. (1969). Yeast genetics. *In* "The Yeasts" (Eds A. H. Rose and J. S. Harrison), Vol. 1. Academic Press, London and New York.

Nasmyth, K. A. and Reed, S. I. (1980). Isolation of genes by complementation in yeast: Molecular cloning of a cell-cycle gene. *Proc. Natn. Acad. Sci. U.S.A.* **77**, 2119–2123.

Nasmyth, K. A. and Tatchell, K. (1980). The structure of transposable yeast mating type loci. *Cell* **19**, 753–764.

Nasmyth, K. A., Tatchell, K., Hall, B. D., Astell, C. and Smith, M. (1981). A position effect in the control of transcription at yeast mating type loci. *Nature, Lond.* **289**, 244–250.

Nelson, R. G. and Fangman, W. L. (1979). Nucleosome organisation of the yeast 2μm DNA plasmid: a eukaryotic minichromosome. *Proc. Natn. Acad. Sci. U.S.A.* **76**, 6515–6519.

Newlon, C. S. and Burke, W. (1980). 9th Annual ICN-UCLA Symposia on Mechanistic Studies of DNA Replication and Genetic Recombination.

Petes, T. D. (1980a). Unequal meiotic recombination within tandem arrays of yeast ribosomal DNA genes. *Cell* **19**, 765–774.

Petes, T. D. (1980b). Molecular genetics of yeast. *A. Rev. Biochem.* **49**, 845–876.

Petes, T. D. and Williamson, D. H. (1975). Replicating circular DNA molecules in yeast. *Cell* **4**, 249–253.

Phaff, H. J. (1971). Structure and biosynthesis of the yeast cell envelope. *In* "The Yeasts" pp. 135–270. (Eds A. H. Rose and J. S. Harrison), Vol. 2, 135–270. Academic Press, London and New York.

Ratzkin, B. and Carbon, J. (1977). Functional expression of a cloned yeast gene in *E. coli. Proc. Natn. Acad. Sci. U.S.A.* **74**, 487–491.

Rose, A. H. and Harrison, J. S. (Eds) (1969). Biology of Yeasts. *In* "The Yeasts" Vol. 1. Academic Press, London and New York.

Scherer, S. and Davis, R. W. (1979). Replacement of chromosome segments with altered DNA sequences constructed *in vitro. Proc. Natn. Acad. Sci. U.S.A.* **76**, 4951–4955.

Van Solingen, P. and Van Der Plaat, J. B. (1977). Fusion of yeast spheroplasts. *J. Bacteriol.* **130**, 946–947.

Stevens, B. J. and Moustacchi, E. (1971). ADN satellite γ et molecules circulaires torsadees de petite taille chez la levure *Saccharomyces cerevisiae. Exp. Cell Res.* **64**, 259–266.

Stinchcomb, D. T., Struhl, K. and Davis, R. W. (1979). Isolation and characterisation of a yeast chromosomal replicator. *Nature, Lond.* **282**, 39–43.

Stinchcomb, D. T., Thomas, M., Kelly, J., Selker, E. and Davis, R. W. (1980). Eukaryotic DNA segments capable of autonomous replication in yeast. *Proc. Natn. Acad. Sci. U.S.A.* **77**, 4559–4563.

Stinchcomb, D. T., Mann, C., Thomas, M. and Davies, R. W. (1981). ICN-UCLA Symp. in press.

Strathern, J. N., Jone, E. W. and Broach, J. R. (Eds) (1981). "The Molecular Biology of the Yeast *Saccharomyces*." Cold Spring Harbor Laboratory, New York.

Struhl, K., Cameron, J. R. and Davis, R. W. (1976). Functional genetic expression of eukaryotic DNA in *Escherichia coli. Proc. Natn. Acad. Sci. U.S.A.* **73**, 1471–1475.

Struhl, K., Stinchcomb, D. T., Scherer, S. and Davis, R. W. (1979). High frequency transformation of yeast: Autonomous replication of hybrid DNA molecules. *Proc. Natn. Acad. Sci. U.S.A.* **76**, 1035–1039.

Szostak, J. W. and Wu, R. (1980). Unequal crossing over in the ribosomal DNA of *Saccharomyces cerevisiae. Nature, Lond.* **284**, 426–430.

Twigg, A. J. and Sherratt, D. (1980). Trans-complementable copy-number mutants of plasmid colE1. *Nature, Lond.* **283**, 216–218.

Walz, A., Ratzkin, B. and Carbon, J. (1978). Control of expression of a cloned yeast (*Saccharomyces cerevisiae*) gene (*trp*5) by a bacterial insertion element (1S2). *Proc. Natn. Acad. Sci. U.S.A.* **75**, 6172–6176.

Zakian, V. A., Brewer, B. J. and Fangman, W. L. (1979). Replication of each copy of the yeast 2 micron DNA plasmid occurs during the S phase. *Cell* **17**, 923–934.

Zeman, L. and Lusena, C. V. (1974). Closed circular DNA associated with yeast mitochondria. *FEBS Letters* **38**, 171–174.